BEHAVIOR PATTERNS OF HYDRATION

by

I0462139

LEWIS H. FLINT
Louisiana State University
Baton Rouge, Louisiana, U.S.A.

edited by Dr. Lokesh Chandra
Institute for the Advancement of Science and Culture
J22, Haus Khas Enclave, NEW DELHI 16 (INDIA)

Foreword by
STUART HALE SHAKMAN
Institute of Science
Santa Monica, CA, U.S.A.

i

PREFACE
Author's Note

This manuscript documents a breakthrough in the important area in which physics, chemistry and biology meet: the area of aqueous solutions. Unexpectedly the breakthrough, as embodied in the description of periodic hydrational potentiality, yielded an explanation of the origin of the system of irregular fractional atomic weight values held in high esteem in contemporary chemistry. Since any revaluation inevitably would become controversial it has been considered necessary, even at the obvious price of repetitive tabulations, to utilize in so important a documentation the safest criterion of integrity known to science : agreement between predicted and observed values. As compensation the treatment extends to the lay reader the unusual experience of sharing in the excitement attending a critical development in science, since this is a first showing, not a rerun. There is here the drama of entering, however slowly and with faltering steps, a new phase of understanding having an intimate relation to life. Retrospectively there was earth, air, water and fire, — but these were not enough, Beyond these the mysteries were allocated to supernatural forces. Yet herewith enter new potentialities for natural forces, and even what we call beauty becomes revealed as a natural outgrowth of the behavior patterns of atoms, — patterns conferring the rewarding satisfaction inherent in a recognition of basic order.

iii

LIST OF TABLES

vii

Foreword – S. H. Shakman

The importance of Lewis Herrick Flint's work on hydration cannot be overstated. It certainly ranks on a par with: the revolutionary works of Aristarchus of Samos, hero of Archimedes and true father of knowledge of our heliocentric universe; Johannes Kepler, whose 3rd harmonic law in particular demonstrated with mathematical precision the algebraic structure of our heliocentric universe; possibly Henry G. J. Moseley, whose researches established a precise algebraic relation between atomic numbers and wavelengths of vibration of the various elements; and precious few others. Of course, the world is familiar with the likes of Archimedes, Copernicus, Newton and Einstein, but even aura of these familiar names arguably must acknowledge the primary true importance of Aristarchus, Kepler and Flint within the grand history of science..

As with the earlier giants (indeed what makes them giants), Flint's work is not theoretical; rather, it is calculated from incontrovertible physical phenomena. Flint's work does not provide all the answers to the mysteries of science that remain, but it does provide a framework and solid foundation that seemingly can and will provide the answers (if we don't self-destroy first).

Simply stated, what Flint discovered was a simple algebraic relationship (inverse and integral) that exists between the atomic number (plus valence) of a given ion, and the maximum and commonly-encountered numbers of water units that are associated with it when that ion is dissolved in water. This entity is commonly referred to as the hydration number, or respective hydration numbers, and their identities and underlying behavior remain among the greatest of mysteries of science.

The importance of the relation between water and things dissolved in or combined with it is immediately evident in the indispensable role of water in all of the physical sciences, prominently including chemistry, physics and biology. Thus even from a cursory glance it is clear that a true understanding of the interaction between water units and its varied associates is indeed a grand unifying scientific principle. Conversely, were the object of scientific inquiry to be the discovery of a "grand unification" scheme, the unique and eminently sensible foundation for such a scheme would logically be the disclosure of the mechanism of interaction between water units and its associates.

As disclosed by Flint, and more fully discussed in this volume, a subsequent volume (*Hydration and Biology*), and Shakman's *Principles of Hydration*:

Table 1. Flint's Description of Hydrational Potentiality

H (the (maximum) hydration number) = 23n – (Z+C);
when Z = atomic number; C = valence; and
n=1 for (Z+C)=0-23,
n=2 for (Z+C)=23-46,
n=3 for (Z+C)=46-69,
n=4 for (Z+C)=69-92; and within each of these periods, the maximum hydration number (H) decreases from 23 to zero.

Figure 1. The Helical Structure of Hydrational Periodicity
        As Z+C increases: 0-23, 23-46, 46-69, 69-92

Z   92
+   ↑   80
C   81
    69 ↑   57
                                                                58
    46 ↑   34
                                                                35
    23 ↑   11
                                                                12
     0 ↑   ←

H decreases from 23 to 0, repetitively.

                21 20 19 18   17 16 15   14
           22   →   13   ↓
H: ↑ 23   12 ↓
    ↑ 0   11
        1   ←   10
             2   3   4   5   6   7   8   9

With the benefit of hindsight, we may illustrate Flint's discovery with direct calculations from well-established non-controversial fundamental data. The data set is comprised of "limiting ionic conductivities", i.e., the relative amounts of electrical current conveyed by the various ions in solution. A tremendous amount of research energy was expended in this area particularly around the turn of centuries, from the 19th to the 20th. Prominent in these researches are found the great names of science of the era, van't Hoff, Nernst, Arrhenius, Kohlrausch and others, including the particularly exhaustive efforts of Harry Jones of the Carnegie Foundation. In this quest, this particular line of inquiry was indispensably enabled by Kohlrausch, whose law of independent migration of ions enabled conductivities of the individual ions to be separated out from the measured conductance of various solutes (e.g., NaCl, KCl, etc.), and isolated for examination in the form of "limiting ionic conductivities" (e.g., Na+, K+, etc). At the same time, these values are also known to represent the relative mobilities, or velocities, of the given ions.

The bases for this illustration of direct calculation of hydration numbers, from limiting ionic conductivities, are the same as used by Flint in his discovery. The fundamental principles are
(1) van't Hoff's decisive demonstration of the analogy between gaseous (atmospheric pressure) and solute (osmotic pressure) behavior [for which van't Hoff was awarded the first Nobel], and
(2) the principle formerly known as Graham's law (now implicit within the law of kinetic energy), whereby gaseous velocity into a vacuum varies inversely with the square-root of the mass – in other words, a gas with 4 times the mass will travel half as fast, with 9 times the mass, 1/3 as fast, etc.

While these two principles were well-established by the time of Flint (1932), it does not appear that anyone prior to Flint had attempted to use these in conjunction with ionic conductivities/ mobilities to attempt to determine relative hydration numbers. While Flint was apparently working from, and adjusting numbers of, Bousfield that were cited in Flint's graduate school Bayliss physiology text (see Flint 1932 and Shakman's *Principles of Hydration* 2014 for further discussion), herein, in Table 2 and Figure 2, is an illustration of Flint's discovery directly calculated from conductivity data kindly provided through referral in 1996 by Prof. Howard Reiss of UCLA to the then-current textbook, Noggle, J.H., *Physical Chemistry*, 1996, p. 411.

Direct Calculation of Hydration Numbers (This direct calculation method had been proposed but not published, registered as *Nature SXA011)*)

There are three steps involved in the calculation of hydration numbers from equivalent ionic conductivities:
(a) total (relative) ionic weights are derived as the inverse-square of respective conductivities, and adjusted relative to the value of a base ion assumed to be anhydrous;
(b) weights of anhydrous solutes are subtracted from total ionic weights to derive weights of water of hydration associated with respective ions;
(c) weights of water of hydration are divided by weight per water unit (18) to derive numbers of water units (hydration number) associated with respective ions.

Equation used in calculations, Table 2 and Figure 2:
Hcalc (calculated Hydration number =
$(k/(\text{conductance}^2) - AW)/18$; k (constant in calculations) $= 517336$

This constant corrects all values relative to an atomic weight of 85.4768 for the "base" ion Rb+, which for the purpose of this set of calculations is assumed to be anhydrous, i.e., to have a hydration number of zero.

Figure 2 plots calculated hydration numbers against respective sums of atomic number and valence. The inverse linear result shown in Figure 1 illustrates the essential foundation of the methodology first encountered and discussed by L. H. Flint in 1932[3], wherein, for the hydrated lighter ions being studied, i.e., Li+, Na+ and K+, the sums of Z+C+H were found to equal 23.

As shown in Table 2 and Figure 2, calculations for H+ and OH-, as well as the base ion, Rb+, also yield approximate linear results at H = 0. The suggestion that relatively large conductivities of H+ and OH- indicated they were not hydrated was first made by Abegg and Bodlander[4] in 1899, first calculated by Flint[3] in 1932 and explained by Flint as evidencing dehydration due to the electrical stress imposed in measuring conductance[5].

1. Noggle, J.H., Physical Chemistry, 1996, p. 411.
2. Gluekauf, E., Faraday Soc., Transactions 51 1241 (1955).

3. Flint, L. H., J. Wash. Acad. of Sci. 22, 97-119, 211-217 & 233-237 (1932).
4. Abegg and Bodlander, Zeit. f. Anorg. Chem. 454-499 (1899).
5. Flint, L.H., Dissenting Ape, Dahlia Street, New York, 1973.2

Table 2 / Figure 2: Inverse square of limiting ionic conductance (l) adjusted to base weight of 85.47 for Rb+; minus respective atomic weights (A.W.), equals weight of water of hydration, divided by 18 equals calculated hydration number (Hcalc).

Calculated hydration number (Hcalc) from conductance (EC) & atomic weight (AW)

Hcalc= ((k/EC-square)-AW)/18; k=517323 when base is Rb+ with Hcalc=0.

Copyr.1999 SHShakman

ION Rb	Z	C	AW	EC	IW calc	WW	H calc	
(BASE)	37	1	85.4678	77.8	85.46781	1E-05	0	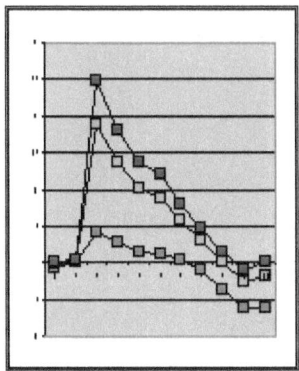
H	1	1	1.0078	349.8	4.227875	3.2201	0.1789	
OH	9	1	17.0073	197.6	13.24915	-3.758	-0.209	
Li	3	1	6.941	38.66	346.1291	339.19	18.844	
Na	11	1	22.9898	50.08	206.2686	183.28	1.182	
Mg	12	2	24.305	53.06	183.75	159.44	8.8581	
Al	13	3	26.9815	63	130.3409	103.36	5.7422	
K	19	1	39.0983	73.48	95.81279	56.714	3.1508	

Table 3/ Figure3: The middle (Rb+) column of Table 3 duplicates the values for Hcalc in Table 2 for a larger range of ions. The other columns of Table 3 use H+ and La+ as base ions (respectively assumed to be anhydrous, as shown) for comparison:

Input-AW (Atomic Wt)

BASE=	H+	Rb+	La+3
OH-	-0.77	-0.21	0.01
H+	0	0.17	0.25
Li+	4.19	18.84	24.7
Be++	2.88	13.69	18.01
Na+	1.45	10.18	13.67
Mg++	1.08	8.86	11.97
Al+3	0.23	5.74	7.95
K+	-0.9	3.15	4.77
Rb+	-3.62	0	1.45
Cs+	-6.24	-2.57	-1.11
La+3	-6.31	-1.8	0

Table 4/ Figure 4: duplicates Table 3 / Figure 3, except for the substitution of Atomic Number (Z) in the place of Atomic Weight (AW):

Input-Z (Atomic   No.)

BASE =	H+	Rb+	La+3
OH-	-0.59	-0.32	-0.19
H+	0	0.08	0.13
Li+	7.87	14.68	18.23
Be++	5.63	10.69	14.74
Na+	3.77	7.83	9.94
Mg++	3.14	6.75	9.59
Al+3	1.78	4.34	6.31
K+	0.36	2.25	3.23
Rb+	-1.68	0	0.87
Cs+	-3.45	-1.75	-0.87
La+3	-3.18	-1.09	0

Table 5/ Figure 5: duplicates Table 4 / Figure 4, except for the substitution of Atomic Number plus Valence (Z+C) in the place of Atomic Number (Z):

Input-Z (Atomic   Number)   +C(Valence)

BASE =	H+	Rb+	La+3
OH-	-0.19	-0.23	-0.21
H+	0	-0.01	0.04
Li+	17.73	16.65	21.22
Be++	12.75	11.95	15.33
Na+	9.5	8.86	11.58
Mg++	8.09	7.52	9.95
Al+3	5.07	4.66	6.38
K+	2.81	2.51	3.78
Rb+	0.27	0	1.13
Cs+	-1.67	-1.95	-0.8
La+3	-1.07	-1.41	0

Table 6 / Figure 6 illustrates the concept of periods, based on calculations of hydration numbers from conductance, as proposed by Flint; for the range of ions shown, the totals of Z+C+H tend to cluster around values indicating the first, second or third hydrational period:

ION	Z	C	H	N=(Z+C+H)/23
Na=base	11	1	11	1
Li	3	1	20.25	1.054413
Be	4	2	14.61	0.896019
Na (base)	11	1	11	1
Mg	12	2	9.431	1.018753
Al	13	3	6.016	0.957202
K	19	1	3.507	1.022029
Sr	38	2	4.322	1.927055
Y	39	3	3.38	1.973051
N3	21	-1	4.275	1.055424
HS	17	-1	5.543	0.936671
HCO3	31	-1	12.29	1.838563
H2PO2	33	-1	11.06	1.872288
HPO4	48	-2	23.29	3.012739
H2PO4	49	-1	23.07	3.090034
PO4	47	-3	1.608	1.98296
CNO	21	-1	5.19	1.095214
SeCN	47	-1	2.278	2.09905
PO3F	48	-1	2.498	2.152065
PF6	69	-1	1.998	3.04341

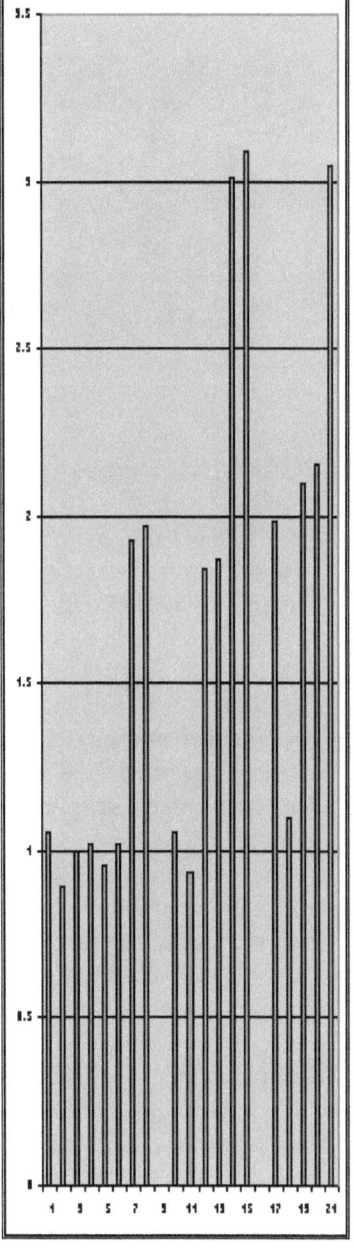

Table 7 below illustrates calculations of both diameters and conductivities for the set of eight ions listed by Ganong.* For a majority (five) of these eight (hydrated) ions, both diameter* & conductivity** may be approximated as cube-root of volume & inverse-square-root of weight, respectively when these ions are assumed to be fully hydrated as per Flint***.

```
IONIC SIZE AND CONDUCTIVITY
Ion       Diameter   Conductivity
           Obs* Calc Ob** Calc H
Na+ BASE 1.47 1.47 50.1 50.1 11
K+          1.00 1.02 73.5 76.9 3
HCO3-       1.65 1.66 44.5 40.0 16
CH3COO-     1.80 1.68 40.9 40.0 16
H2PO4-      2.04 1.84 33   34.3 21
Cl-          .96 1.28 76.3 59.4 7
" alt.       .96 1.04 76.3 76.6 3.5  *
HPO4--      2.58 1.89 33   33.2 23
" alt.      2.58 2.39                *
SO4--       1.84 1.89 80   33.2 23
" alt.                 80   77.8 0   *
```

*GANONG, W.F., *Review of Medical Physiology* (1975) 12.
** CRC (1985-6) D167-8.
***FLINT, L.H., *Behavior Patterns of Hydration* (1964) 21-30:
Wa[anhydrous weight]=2(atomic # +- valence); Wh[hydrated weight]=Wa+18H; H[hydration#] =23n-(atomic # +-valence) [H=23 to 0, n=1 to 4]; Vh[hydrated volume]=Wh/{1+(Wa/Wh)}.

* Anomolous results may be improved (alt.) for:
(a) both size and conductivity of the Cl- ion by assuming an hydration number [H] of half the prescribed maximum. An hydration number of 3.5 was first suggested by Bousfield, and could represent a type of sharing or bonding;
(b) the diameter of HPO4-- by speculatively deriving it as the cube-root of 2 fully-hydrated ionic volumes, while this ion calculates as a single fully-hydrated ion when measured for conductivity. This could be taken to suggest that this ion, when not under electrical stress, may exist in a binary form, as do many gaseous atoms, e.g., H2, O2, N2, etc.; but is split into its two parts when under electrical stress, while each part holds its water of hydration complement.
(c) conductivity of the SO4—ion by assuming it is fully hydrated when measured for diameter, and anhydrous when measured for conductivity. The precise data suggest that the sulfate ion releases all of its water of hydration

under electrical stress, becoming an anhydrous ion. The potential of exploiting such an apparent electrical ionic switching mechanism in biological / medical situations seems worthy of wistful contemplation.

Dr. Flint discusses in depth variations in hydrational status as a result of electrical stress or other conditions, such as hydrational bonding, in his two volumes, usually in a physiological context. The importance of these concepts to animals such as humans is self-evident, for without the likes of hydrational bonding we would collapse in a puddle of mostly water. Beyond this all-encompassing fact, the reality of life's processes being governed by eminently calculable (and presumably thus reversible) osmotic processes presents the exciting prospect of inevitable control of these processes, e.g., even those of disease and aging. The prospect of truly understanding and managing transitions between matter and energy, not merely on the explosive scale of nuclear energy but on the internal living scale, the transition between matter consumed and energy expended in the individual, in a precise and algebraically calculable form, is mind-boggling. But the significance is by no means limited to biology.

The calculations above and further mathematical explorations set out in *Principles of Hydration* are offered as a measure of potential validation, but are not suggested as definitive. Indeed, Flint himself has emphasized that he is outlining and extending his discovered methodology as well as he was able, but that nothing was sacrosanct and beyond improvement. Indeed what we have here is a foundation and framework for further exploration, study, and wherever feasible, implementation.

Flint's laws of hydration are merely a tool for unlocking the mysteries of science, but arguably the most powerful and all-encompassing tool ever encountered. Exposure to this magnificent body of work, and the opportunity to possibly extend it and be a vehicle for helping share it with the world, are certainly among the most thrilling, gratifying and humbling experiences that anyone might enjoy in a precious, brutally short, lifetime.

Stuart Hale Shakman
Santa Monica, CA, USA
August 8, 2014

Lewis Herrick Flint

circa 1917-1918   circa 1960

BEHAVIOR PATTERNS

OF

HYDRATION

by

LEWIS H. FLINT

INTRODUCTION

The release of nuclear energy in substantial measure serves to characterize our era as an explosive one. The population is exploding, the ambitions of the underprivileged are exploding, scientific research is exploding. On the frontiers of research in physics explosions have produced and disclosed a confusing multiplicity of sub-atomic entities, for the most part in complementary pairs of units, a body and an antibody. The composition of atoms naturally is of great interest to a chemical science whose province long has been the behavior of atoms, and in such areas as radioactivity and spectroscopy physics and chemistry have become integrated. Yet in such areas as well as in the release of sub-atomic entities unusual stresses have been involved compared with most phenomena involved in the protoplasmic metabolism which mediates life. Unusual stresses also have been involved in solar radiation, in electrical storms and in thousands of gadgets which have become familiar adjuncts of civilization.

It is inevitable that in the midst of stress and confusion there should arise increasingly a keener, deeper appreciation of pathways to a more orderly and tranquil world. In sharp contrast to unusual stresses appraised as explosive sources of confusion there is a background of less-excited atoms whose behavior is responsible for the more peaceful phenomena of natural science. It is this behavior which is the primary concern of this book. The less-excited atoms are the chemical elements at energy levels below those involving radiation. The behavior patterns of these elements have engendered the development of chemical science, a science whose contributions to human welfare are widely recognized and appreciated. Quite naturally and appropriately the major emphasis of chemical science has been placed upon the behavior patterns of these elements *with respect to each other*.

This book describes results obtained in researches on the behavior patterns of the chemical elements not so much with respect to each other as *with respect to an aqueous solvent*. In relation to conventional chemistry the results appear subject to appraisal as interpretive and supplementary. They seem to be of special significance in relation to processes in the areas of biophysics and biochemistry. In substantial measure they supply a satisfying explanation of the benefaction of rain to a plant, of a drink of water to an animal. For an explosive and confusing era they mark a pathway to atoms with inherent mathematical attributes and assured potentialities for playing quiet roles in predictable patterns of behavior.

CHAPTER 1

ORIENTATION

For some time there has been the distracting notion that I should interrupt my absorbingly pleasurable researches long enough to record some words about them and to discuss their results. Here is my interruption. It has to be in the area of autobiography for my researches began in a period before togetherness had entered the laboratory and they have continued to this moment in the ivy tower tradition. The consequent invasion of the impersonal by the personal departs from the severe front of objectivity fostered by contemporary scientists. Being human is great fun, and being honest as well as curious is the essence of scientific endeavor.

The context of this book, if I ever manage to complete it to my own satisfaction not to mention that of anyone else, will report on researches which have extended over a period of about thirty years. During this period such things as jets, missiles and satellites have imposed a nervous impetuosity on many people. Fast everything has extended to include fast reading,—but at least for some people it takes time for some thoughts to sink in. There will be many such thoughts in the chapters to follow. Years ago in the nation's capital a restaurant menu greeted its prospective patrons with some such caption as "It takes time to prepare food in the manner in which we are proud to serve it. If you are in a hurry we suggest you go elsewhere." It was a pleasant thought to remember for an age of hurry. The cult of human scanners will not attain understanding here. As for mysteries, those in this book will be impersonal and will relate to the behavior patterns of such invisible and miniature ghosts as atoms and molecules. The mysteries are frustrating because they keep introducing other mysteries until ignorance and bliss really do seem to be correlated. Yet we know that is not true: the bliss of ignorance is imaginary and the things that have been thus imagined have nothing of appeal in comparison with the wonders disclosed by knowledge. The bliss of ignorance is the easy lazy approach. The bliss of knowledge is the difficult approach: the many mysteries are irresistibly fascinating only in the event that one has the time, patience, urge and mental equipment to accept their challenge. Under such conditions the search for knowledge can be an introduction to a sort of paradise: few thrills are greater than those of a scientist at the advent of discovery. That explains why the search for knowledge is so all-absorbing.

There are many potential adjuncts to a scientist's enjoyment of his paradise and one is a keen delight in nature. I have slept above a timberline and have awakened to watch the sun appear above an ocean of fleecy white clouds below. In these experiences alone there is something of the joy of discovery: the possession of light that has not reached the inhabitants of the still-darkened villages in the valleys below. In science, however, the discovery,—the possession of knowledge that has not reached others—often may initiate a chain of reactions, disclosing like the turns of a kaleidoscope new patterns for seemingly imperative investigations. Thus in science one may yield to the calls for additional research and delay or ignore any recognition of a subtle sense of obligation to share knowledge. On occasion such a sequence of events may not be devoid of advantage, for incident to publication one may meet devout guardians of conformity quite capable of impairing zest for further research. Insofar as the considerations within this book are involved my first important discovery took place in Washington, D. C., in 1932 in what was then the Bureau of Plant Industry of the United States Department of Agriculture. The discovery was made in an environment of researches on the electrical conductivity of aqueous solutions, or electrical resistance, if one prefers. It is possible that in some darkened valleys there might be found even today some villages unaware of the fact that the development of the Wheatstone bridge brought to electrical conductivity measurements a refinement conferring sensitivity approaching the seemingly occult. Many a student has felt reasonably clean or pure before taking a course in bacteriology, only to become aware in course of an intimate harboring of countless numbers of bacteria. In a similar way the Wheatstone bridge mediated the disclosure of impurities in water which otherwise might have been appraised as clean or pure.

The important discovery, like most discoveries, came about through a sort of confluence of curiosity and accident. The curiosity concerned the significance of electrical conductivity data for aqueous solutions in general and the relative mobility of specific solute ions in particular. A substantial accumulation of data had been obtained by various scientists over a period of about fifty years. The analogy between solute units and gases had been memorialized in the Kohlrausch law of the independent migration of ions. The Graham law of diffusion had been recognized for more than fifty years. The curiosity was simple: why could not the electrical conductivity data and the principles be integrated? The answer was not forthcoming until the accident.

The accident involved in the important discovery was the chance finding of an article published in Germany in 1899 and written by ABEGG and BODLÄNDER. As the author of the valence eight rule ABEGG attained something of a hold on immortality in the annals of chemistry, but the mentioned article apparently failed to elicit notable attention. Perhaps it was on this account that the involved principle was not extended in such a manner as to disclose its periodic nature. The article by ABEGG and BODLÄNDER was entitled "Electro-Affinity—A New Force in Chemistry." The force was not really new: several earlier publications had been concerned with it. What appeared to be new was the recognition of the inverse relationship between the force and ionic weight. There seems no reason to doubt that ABEGG possessed an adequate background for extending the principle to its logical conclusion. One may venture to surmise, therefore, that since the extension involved aspects controversial to contemporary viewpoints in chemical science, he may have lacked the courage. So natural is human addiction to long-accustomed patterns of thought that often there exists a practically instinctive aversion to change. So natural also is human desire for the tolerance, if not the esteem of others, that often there exists a restraining reluctance to give expression to concepts certain to meet deprecation. The situation is but further evidence of the inevitable commingling of the personal and the impersonal in science.

The force involved in the article by ABEGG and BODLÄNDER was the force which mediates hydration. Then, as now, it could be designated as an electro-affinity; but clearly it did not involve the sharing of electrons which has come to characterize the type of bondage typically present in chemical compounds. It was a force which, although readily demonstrable, in substantial measure seemed to challenge categorical description. Based primarily on the observed behavior of the solute ions Li^+, Na^+, and K^+, the force was described as inversely related to ionic weight and nothing further was attempted. So near and yet so far were ABEGG and BODLÄNDER from making a most important discovery. There seems to be no question but that they were familiar with the Graham gas law of diffusion since their publication offered "electro-affinity" as an explanation of the obvious failure of the Kohlrausch element ion mobilities to comply with the calculated or expected ion diffusion mobilities. Under the Graham law the mobilities would be projected as inversely proportional to the square roots of the involved weights. Thus in the Li^+ - Na^+ - K^+

series the Li^+ ion would have the greatest mobility and the K^+ ion the least. The Kohlrausch electrometric data, however, yielded ion mobilities in the reverse order: that for the K^+ ion was the greatest, that for the Li^+ ion was the least. The important contribution of ABEGG and BODLÄNDER was the contention that of the three ions the Li^+ ion had the greatest ability to take on water of hydration and the K^+ ion had the least such ability. In aqueous solution and under electrical stress, therefore, the ions of the Li^+ - Na^+ - K^+ series were interpreted as having become hydrated in a decreasing order so that in reality the weight of the hydrated Li^+ ion was greater than that of the hydrated K^+ ion, the weight of the hydrated Na^+ ion being considered as intermediate. As thus represented the Kohlrausch data supplied mobility values for hydrated ions whereas the Graham law had supplied mobility values for anhydrous ions. From this viewpoint it was obvious that the electrometric data supplied indices of the relative weights of the hydrated ions through correlation with the Graham law and that the relative weights of the hydrated ions would have significance with respect to the numbers of H_2O units held in hydration. One may surmise, therefore, that with even a little encouragement by contemporary chemists ABEGG almost certainly would have followed up the "new force" and would have discovered that the relationship between ionic weight and hydrational potentiality was not only inverse as stated, but was also integral, reciprocal and periodic.

This was the important discovery which followed the accidental finding of ABEGG and BODLÄNDER's article in 1932: that the relationship between ionic weight and hydrational potentiality was inverse, integral, reciprocal and periodic. Individually these adjectives are applicable to many common mathematical appraisals of phenomena. Collectively no other phenomenon in nature is known to which they would all be applicable. On this account the relationship between ionic weight and hydrational potentiality is here considered as unique.

It may be taken for granted, without specific admission or comment that the mentioned important discovery took place without the involvement of any extraordinary mental prowess. The first step consisted in solving the simple mathematical problem of determining the minimal numbers of water molecules of weight 18 which, when added to weight values representing Li^+, Na^+ and K^+ in the anhydrous states, would yield total weight values whose diffusion mobilities under the Graham law would approximate the observed mobilities under electrical stress, all values being compared on a relative basis. The involved calculations were quite simple with the aid of a suitable slide rule. It is to be recognized that since the procedure was empirical with numerous postulates the attainment of reasonable approximations without such an aid would have been a tedious time-consuming task unless one was particularly adept in mathematics. Whether or not one may account for the failure there seems to be no evidence that the Graham diffusion law had ever been used previously in conjunction with the principle of ABEGG and BODLÄNDER.

The values obtained in the indicated first step might be described as representing a near approach to mathematical integrity. It was quite clear that further empirical postulates were in order, and since the numbers of H_2O molecules used in the first step were integers it followed that the direction of improved correlation concerned the irregular fractional atomic weight values for the Li^+, Na^+ and K^+ ions. For the second step, therefore, the respective ionic weight values were postulated as twice the atomic numbers. This procedure yielded an improved order of correlation, almost but not quite satisfactory. The second step obviously involved a transgression from abject faith in the integrity of the atomic weight values for the specified ions, and this was a conditioning factor in the subsequent search for a more perfect correlation. In the meantime it had become obvious that although the Li^+, Na^+ and K^+ ions were being used as key index ions any correct interpretation of their patterns of behavior under electrical stress well might be expected to have significance over the extensive range of phenomena involving solute ions is an aqueous solvent. This realization imposed an enhanced aura of conservatism and restraint upon ventured postulates: nothing short of mathematical nicely in correlation and integration could safely be tolerated. Yet actually the corner stone of mathematical integrity had been laid in the closing decades of the previous century by the deservedly renowned meticulous analytical researches of chemists. These researches very definitely established the prevailing incidence of whole numbers of H_2O units per molecule of solute when solutes crystallized out of solutions under conditions yielding hydrated crystals. The results documented the integral aspect of hydrational potentiality, even though the documentation was something of an inadvertent by-product of the utmost painstaking in chemical analysis.

Although the substitution of whole numbers for fractional atomic weight values to enhance correlations with integral H_2O units was inherently provocative within the area of what might be termed orthodox chemical science there was an extenuating aspect for consideration. This was the fact that with a few notable exceptions the atomic weight values of the chemical elements had been obtained from measurements involving the solid state, whereas the electrometric data involved solute units with an individuality analogous to that characteristic of gases. In such notable exceptions as helium, carbon, nitrogen and oxygen the derived atomic weight values had been integers.

The third and final step which marked the definitive diagnosis of hydrational potentiality was the introduction of the postulate of a specific change in weight with ionization. This change was oriented to a base value of twice the atomic number of the element, appraised as the weight of the element in the neutral state, with the element helium as the outstanding example. From this neutral state the weight was projected and then evidenced as increased by 2 for each positive charge and decreased by 2 for each negative charge. It was a drastic revolutionary and seemingly incredible step,—but it permitted an integration of observational data with a satisfying convincing nicety. It led to the previously indicated characterization of the hydrational potentiality of ions as an inverse, integral, reciprocal and periodic relationship, unique in nature. All of the succeeding chapters of this book directly or indirectly will be concerned with data evidencing the integrity of the relationship.

At the time of the discovery in 1932 I had attained the age of 39, an age held and maintained by no less an authority than Jack Benny to represent the very peak of perfection in a human male. Under the circumstances there naturally was engendered a measure of personal pride and enthusiasm, and basic mathematical data for weight, hydration and mobility covering the four periods were prepared. These data were published in the Journal of the Washington Academy of Sciences during that year. With respect to these data it might be said that history repeated itself and that the publication of the extension of Abegg and Bodländer's principle was received in a mánner similar to that of their original paper. The situation is reported merely as factual and without any implication of complaint. In scientific research patience and perseverance are engendered by a negative response to positive results. Most of the chapters of this book will deal with researches directly or indirectly prompted by the negative responses and a sustained confidence in the positive results.

At the present time it will be recognized by many that during the past thirty years a change has taken place in chemical science. During recent years it has been realized increasingly by many chemists that hydration is an intimate adjunct of the behavior patterns of solutes in aqueous solution. The development has been an interesting one to watch from the ivivied tower; for human ingenuity in explaining situations not understood may invoke the supernatural in science as it so often has done and continues to do in religion. Without an adequate description of hydrational potentiality floundering is inevitable. Moreover, with the present surfeit of chemical literature it has seemed reasonable to venture that the 1932 description of hydration has been successfully buried. It will be ressurrected in the following chapter, wearing mathematical clothes. You may not like it: it is not complimentary to chemists who have been intensely concentrating their attention on the behavior of solute atoms and molecules with respect to each other and who have inadequately considered the behavior with respect to their aqueous solvent. You may not like it: it supplies a basis for an alternative interpretation of the extensive gravimetric data supporting the contemporary incongruous aggregation of irregular, fractional and believe-it-or-not largely fictitious atomic weight values. You may not like it: it introduces the attribute of counter-balance potentiality in such a manner as to involve the difficult-to-scan concept of anti-gravity or negative weight. You may not like it: it embodies a periodicity quite different from the periodicity after Mendeleev, now a proud possession of chemical science, but regrettably a subject of derision by chemists at the time of its introduction. You may not like it, but science has yet to discover a way to advance the pleasure of some without including a measure of dismay for others. You may not like it, and you may not believe it, but the description of hydrational potentiality as given and validated in the chapters of this book is a doorway to a recognition and appreciation of a more beautifully ordered nature by far than anything it has hitherto been justifiable for you to imagine.

LITERATURE CITED

1. ABEGG, R. & G. BODLÄNDER (1899). Die Elektroaffinitat, ein neues Prinzip der Chemischen Systematik. Zeitschr. Anorg. Chem. 20:453-499.
2. FLINT, L. H. (1932). Hydration of the solute ions of the lighter-elements. Jour. Wash. Acad. Sci. 22:97-119.
3. —(1932). Hydration of the solute ions of the heavier elements. Jour. Wash. Acad. Sci .22:211-217.
4. GRAHAM, T. (1866). Elements of Inorganic Chemistry. 2nd Amer. edition. pp. 88-89.
5. KOHLRAUSCH, F., & H. VON STEINWEHR (1902). Weitere untersuchungen über das Leitvermogen von Elektrolyden aus einwerthigen Ionen in Wasseriger Losung. Akad. d. Wiss. Berlin Sitzungsberichte. pp. 581-587.

CHAPTER 2

PERIODIC HYDRATIONAL POTENTIALITY

The key value implementing the practical evaluation of the hydrational force operative in ions of a specified weight range is the number of H_2O units of weight 18 which maximally can be held by the ion. This key value has been assigned the symbol H. In mathematical terms periodic hydrational potentiality may then be represented by the formula

$$H = 23n-(A.N. \pm C)$$

In which A. N. = the atomic number of an element, in the case of an element ion, or in the case of molecular ions the sum of the atomic numbers of each component atom; C=the valence of the ion, this number to be added if positive, subtracted if negative; n=an integer representing the period: to the base $O_2 = 32$ the values of n are 0-44, n=1; 46 - 90, n=2; 92 - 136, n=3; 138 - 184, n = 4. As thus represented hydrational potentiality is restricted to ions over a weight range commensurate with the 92 naturally-occurring elements.

There are four corollaries of the above-represented description. The corollaries delimit categories of weight. The numbering of the corollaries has been arbitrary and has no relation to relative importance or usage.

Corollary 1 $Wn = 2 A.N.$

    When Wn=the weight of the involved unit in the neutral state. For an element atom the value would be twice the atomic number as indicated. For a molecular unit the value would be the sum of the values for the component atoms.

Corollary 2 $Wa = 2(A.N. \pm C)$

    When Wa=the weight of the involved unit as an anhydrous ion. For an element atom the value would be as indicated. For a molecular unit the sum of the A. N. values for the component atoms would be substituted for A.N. C=valence, this number to be added if positive, subtracted if negative.

Corollary 3
$$Wc = \frac{Wn + Wa}{2}$$

When Wc=the combining weight of the involved anhydrous ionic unit. Other symbols as in foregoing corollaries.

Corollary 4
$$Wh = Wa + 18H$$

When Wh=the weight of the completely hydrated ionic unit.

Wa=as in Corollary 2.

H=as in the formula describing hydrational potentiality.

It may be helpfully clarifying to discuss the four corollaries before taking up the consideration of the description. It may be pointed out that in chemical science "atomic weight" has been a general or generic term devoid of subordinate characterization. Conventional chemical usage, for example, would ascribe an atomic weight of 4 for the helium atom. Corollary 1 also would prescribe an atomic weight of 4 for the helium atom and also prescribe for it a neutral state. Corollary 1 would prescribe a weight of 16 for oxygen atoms in the neutral state: but whereas oxygen atoms are not known to be stable in the neutral state, helium atoms in such a state are well known. One may venture what seems to be a reasonable doubt that any attribute of neutral atomic oxygen has ever been measured. One may venture further that the atomic weight $O = 16$ was inherited from a past era in which atoms were appraised as solid balls of the utmost minuteness and a weight of 32 was assigned to biatomic molecules of gaseous oxygen. Under such conditions the inference $O = 16$ well might emerge as axiomatic. Yet for many years now the atoms have been recognized as anything but solid, as being for the most part vacuous. We have no doubt whatsoever with respect to the validity of $O = 16$ when reference is made to neutral atoms of oxygen, even though we have never been able to obtain such atoms in a state sufficiently stable to permit study. The oxygen atoms we have studied have been bivalent anions and bivalent cations. The bivalent anions have been studied to a greater extent, since they have appeared to dominate in inorganic chemistry. The bivalent cations of atomic oxygen have been found and studied primarily in conjunction with the dissociation of some organic solutes. The prevalence or infrequency of the ionic forms of atomic oxygen will be considered further in subsequent chapters: the important point here is that whenever atomic oxygen has been studied in these researches the bivalent anions have been evidenced as having had a weight of 12 in the anhydrous state and the bivalent cations have been evidenced as having had a weight of 20 in the anhydrous state. These were the respective ionic weights that would be prescribed for the ions by Corollary 2. Oxygen has been cited here as an example because of the important extent to which the O^{--} ion of inorganic chemistry has been involved in the derivation of conventional atomic weight values. The involvement might be construed as tragic, but it was a natural error best corrected and forgotten. The integrity of Corollary 2 has been established over and over again in various aspects of solution phenomena, as will be made clear in the course of subsequent chapters.

Corollary 3 represents an inevitable consequence of Corollaries 1 and 2 when ionization and the correlated change in weight are associated with the gain or loss of electrons and chemical combination involves electron sharing. It will be recognized that combining weight as herewith defined is quite different from combining weight as defined in chemical science. However, it is recognized that the latter usage widely has been abandoned as incongruous. A few conventional atomic weight values approximate the Corollary 3 combining weights, perhaps testifying to a uniform type of ionization. The Li^+, Na^+ and K^+ ions which played so important a role in the development of the description of hydration might be cited as examples, with prescribed combining weights as follows: $Li^+=7$, $Na^+=23$, $K^+=39$.

The categories of weight represented by the first three corollaries pertain to the anhydrous state. The values derived through Corollary 1 on a theoretical basis are applicable to all atoms and molecules. On a practical basis they are particularly applicable at the atomic level to the elements of the rare gas series: helium, neon, argon, krypton, xenon and radon. At the molecular level they are particularly applicable to the neutral gases whose homogeneity made possible the statistically uniform behavior patterns so nicely documented in the gas laws. The values derived through Corollary 2 similarly on a theoretical basis are applicable to all atoms and molecules. On a practical basis they are particularly applicable in two areas, each extensive. One of these is the area of chemical reaction, ionization being prerequisite. Although this is the area with which chemical science has been most intimately involved, the determination of ionic weight has not been an easy or satisfying accomplishment. The closest approach to such a goal in conventional chemistry has been the use of mobility under electrical stress, interpreted in conjunction with the Graham Law of Diffusion.

Researches of this nature involving solute ions were meaningless in the absence of a knowledge of hydration. Researches of this nature involving ionized gases yielded results rendered variable and inconclusive by the failure to effect uniform ionization. These data are of great interest, nevertheless, and will be discussed in a later chapter. The other area in which the values derived through Corollary 2 are of great practical concern is the area of hydration, since hydrational potentiality is conditioned solely by weight and without regard to the composition of the involved ion. On account of this relationship the weight of an ion in the anhydrous state is revealed with great simplicity and nicety by the extent of the hydration taking place under conditions permitting a complete satisfaction and expression of hydrational potentialities. Several chapters in this book will be concerned with procedures for determining the extent of hydration and with data obtained in the respective manner relating to the ionic weights of a considerable number of element ions and molecular ions both in the anhydrous and the hydrated states. The data involving the hydrated state document the integrity of Corollary 4, and these data when sufficiently extensive to fairly represent ionic weight values for the anhydrous state throughout the range represented by the 92 naturally-occurring elements in turn document the integrity of the formulated description of hydrational potentiality. At this point it seems appropriate to direct special attention to the description of hydration as given at the beginning of this chapter.

The various stages in the development of the description of hydrational potentiality may be indicated as follows.

1899		Li^+	$-Na^+$	$-K^+$	Ions
		6.94	22.997	39.1	Weights
			—decreasing—		Hydration
1932	(1)	Li^+	$-Na^+$	$-K^+$	Ions
		6.94	22.997	39.1	Weights
		19	11	3	H_2O Units
	(2)	Li^+	$-Na^+$	$-K^+$	Ions
		6	22	38	Weights $(2 \times A.N.)$
		19	11	3	H_2O Units
	(3)	Li^+	$-Na^+$	$-K^+$	Ions
		8	24	40	Weights $2 \times (A.N.^+C)$
		19	11	3	H_2O^- Units

The third step in the 1932 developments was the crucial one which disclosed the deep-seated and hence far reaching significance of hydrational potentiality. It was noted that for each ion one-half the weight value plus the number of H_2O^- units equalled 23, which number was one-fourth the natural system of 92 elements. From these relationships the inverse, integral, reciprocal and periodic aspects became obvious, as indicated in the following.

A. Li^+ Na^+ Ions
    4- 5 - 6 - 7 - 8 - 9-10-11-12   Ionic Wt $\div$ 2
    19-18-17-16-15-14-13-12-11 H_2O^- Units

B. Li^+ K^+ Ions
    0 - 1 - 2 - 3 - 4   20 - 21 - 22 - 23   Ionic Wt. $\div$ 2
    23 - 22- 21 - 20 - 19   3 - 2 - 1 - 0 H_2O^- Units

C.   0—46—92—138—184   Potential Ions :   Ionic Wt.
    23 23 23 23   0 ⎫
     (0) (0) (0)   (0) ⎬ H_2O^- Units

In scheme A, above, the first portion of 1932, step 3, is expanded to indicate the mathematical correlation on an inverse integral and reciprocal basis. In scheme B, both terminations of the first hydration period are indicated. In scheme C, the overall periodic relation of hydration to ionic weight is indicated.

It was obvious from the foregoing patterns of behavior as projected from the data for the Li^+, Na^+ and K^+ ions that the description of hydrational potentiality by virtue of its mathematical base would possess one of the great assets of any description in science: the potentiality for prediction. On this account it became a simple matter to calculate ionic weights and mobilities in both the anhydrous and hydrated states for any specified ions within the four indicated periods. What proved to be difficult, however, was the satisfactory correlation of these calculated or predicted values with observational data. The mobility values for the Li^+, Na^+ and K^+ ions had been derived from the data of transference measurements in which a common anion had been involved. When data for the specific molecular electrical conductivity of any salt were considered, two disturbing features arose. One of these was the fact that the concentration was predicated on atomic weight values which in most instances were greater than the values prescribed by the description of hydrational potentiality. It might appear that one could readily correct for the use of the contemporary atomic weight values. Such might have been the case whenever solutions at very low concentration were prepared directly, as when slightly soluble salts were involved. The other disturbing feature, however, was the fact that in most instances relatively concentrated solutions were prepared and the electrical conductivity data involved the use of seriate volumetric dilutions from such concentrated solutions. Since hydrational potentiality was not recognized, all of the water in an aqueous solution was considered as solvent: but whenever hydration took place it involved a transfer of solvent to solute, a procedure which automatically increased the concentration. What further complicated the situation was the fact that commonly the supposed concentration of the original stock solution before volumetric dilution was not recorded. Under these circumstances it was possible to establish excellent correlations with provisos which read something like this: if KOHLRAUSCH used the atomic weight values current in 1880 and prepared a stock solution of a specified chemical at 3.0 molar concentration, then by progressive volumetric dilutions he would have attained the value he did, but this would not have been correct because he used the wrong atomic weight values and made no allowance for hydration. To predict a result which a scientist might have obtained in the event that he made a series of errors did not prove to be a satisfying procedure. Subsequently, it became obvious that there were other ways in which the integrity of the description of hydrational potentiality could be established, some of which were far less involved than the use of electrometric data.

In the interim consideration was given to some of the more disturbing features of the description of hydrational potentiality and its corollaries. Perhaps the most disturbing feature was the prescribed and evidenced change in weight with ionization. The classical and meticulous researches of MILLIKAN with oil droplets and electrical charges had yielded data assigning a diminutive but positive weight to an electron. There was the problem, therefore, of explaining how an electron of such a weight could have the power to counteract a weight of 2, many times its own magnitude. Yet it was recognized that some of the behavior patterns of radioactive elements were such as to evidence that the loss of an electron moved the element one place forward in the atomic system, which was the equivalent of adding a weight of 2 on the He=4 basis. Further, some of the observational data on the mobility of ionized gases suggested that negative ionization effected increases in mobility commensurate with a reduction in weight of the order of 2 under the diffusion law. Of more immediate concern, quite naturally, was the correlation of the precept with observational data: the procurement of evidence documenting the validity of the prescribed change in weight with ionization.

Another major disturbing feature of the description of hydrational potentiality was its characterization of hydrogen. The neutral atom of hydrogen had a prescribed weight of 2; the positive ion H^+ a weight of 4 and a combining weight of 3; the negative ion H^- a weight of 0 and a combining weight of 1; the biatomic molecule H_2 a weight of 4. These prescribed values certainly would not readily be accepted by contemporary chemists and yet each had to be valid since otherwise any departure would nullify the basic description and its corollaries. For the apprehensive or unbelieving it may be stated here that in the course of this book all of the above characterizations of hydrogen will be validated with observational data. In numerous instances, moreover, these data were obtained by German chemists of presumed impeccable techniques and unquestioned integrity. On this account the prescribed characterization of hydrogen which for some years was a disturbing feature of the description of hydrational potentiality is now subject to recognition as entirely valid. This development at long last represents hydrogen as the basic and substantial building block for the system of elements, just as PROUT envisioned it to be before the Royal Academy of Sciences in 1815. At times the progress of science seems halting, and the birth of an idea, like the birth of a baby, has aspects of pain as well as of joy. Prout's idea that all of the elements were aggregates of a common unit was not given a royal welcome, and throughout the many intervening years as Prout's hypothesis it has been appraised by chemical science as an idyllic dream without foundation in fact. Yet the truth is,—as will be made evident repeatedly in subsequent chapters—that Prout's hypothesis was not supported by the observational data *as interpreted*, but is well supported by the same data interpreted in conformity with the description of hydrational potentiality and its corollaries. The situation more or less obviously emphasizes another truth: that the utmost refinement in technique of measurement in no way insures corresponding accuracy in interpretation.

As it was with Prout's idea, so also was it with the idea of Newland that the elements, when arranged in the order of increasing atomic weight evidenced a periodicity of attributes and behavior patterns. History is sometimes but not always a good teacher. Impetuous self-assuring chemists of a quick mind to deprecate the description of hydrational potentiality possess a truly choice item for sober contemplation. The idea of Newland as represented in a periodic chart of the elements after Mendeleev more than any other thing now typifies chemical science throughout the civilized world. Chemistry proudly proclaims as its own an idea which at its birth it vehemently rejected.

The description of periodic hydrational potentiality with its four corollaries in reality defines basic order for the chemical elements in whatever state they may occur: gaseous, liquid, solid, anhydrous, hydrated, neutral, ionic, free or combined. It would be abjectly absurd to make such a claim for the description without a wealth of observational data to support it, and for the most part the supporting data, as cited in subsequent chapters, were obtained by chemists or by other scientists equally renowned for painstaking and conscientious attention to detail. Since these data were obtained in the absence of a definition of basic order and hence without a mathematical background from which to derive specific objectives the observed values constitute impressive testimony relating to the integrity of measurement. These measurements, alternatively interpreted, in turn document the integrity of the defined basic order.

CHAPTER 3

SPECIFIC GRAVITY OF AQUEOUS SOLUTIONS AS AN INDEX OF PERIODIC HYDRATIONAL POTENTIALITY

It was inevitable that the formulated description of periodic hydrational potentiality with its corollaries delimiting categories of weight should prove provocative or incredible pending the acquisition of observational data validating its integrity. Certainly any implication of error in relation to the conventional system of atomic weight values should not be given or taken lightly.

The description of hydration originated from a survey of electrometric data and it was considered imperative that the documentation of its integrity should involve data of a different nature. In a survey of the attributes of aqueous solutions it was recognized that such data as pertained to electrical conductivity, freezing point depression and boiling point elevation fell into a class in which the concentrations of the involved solutes commonly were based upon contemporary atomic weight values. This prescribed that any attempt to interpret these data in support of the description of periodic hydrational potentiality would involve "correcting" for the use of the contemporary atomic weight values. To venture such corrections after the description of hydration had been evidenced as valid might be appropriate or permissible, but to do so beforehand seemed undesirable.

In such data as pertained to the specific gravity of aqueous solutions the involved concentrations commonly were expressed as percentages of anhydrous solute by weight or as grams of anhydrous solute per liter. It was held, therefore, that the specific gravity of aqueous solutions represented a type of measurement of potential service in the validation of the description of periodic hydration and its corollaries. This proved to be the case, and fortunately a great many measurements of the specific gravity of aqueous solutions had been made in past years by numerous scientists of reliable repute. The measurements were not all made at the same temperature, and temperature demonstrably influenced the specific gravity of an aqueous solution. The measurements involved numerous concentrations, and the citation of numerous values for numerous solutes obviously represented an impractical procedure. Under the circumstances it seemed best to cite specific gravity data for a representative number of solutes in aqueous solution, for a single concentration, and without regard for a uniform temperature. The data cited were taken from the International Critical Tables, the Smithsonian Tables or from handbooks of chemistry and physics. In numerous instances the same data were to be found in all of these listed sources.

It was recognized that although the description of hydration supplied a procedure for determining on a relative basis the weights of solute ions in the anhydrous and completely hydrated states, it did not provide for the determination of corresponding relative volumes in either the anhydrous or the completely hydrated state. Yet for any analysis of specific gravity data and correlation with the description of hydration it was absolutely necessary that the relation of hydration to volume be ascertained. This project naturally involved empirical procedures, but fortunately the relationship proved to be a very simple one. Expressed as a formula the description was as follows.

$$D = 1.0 + \frac{Wa}{Wh}$$

When D = the density of the solute ion relative to water, Wa = the weight in the anhydrous state, and Wh = the weight in the completely hydrated state.

In use the above description was applied in the following manner. All observational data on the specific gravity of aqueous solutions which did not involve concentrations expressed in terms of grams of solute per liter were transposed to that basis. In most instances "grams per liter" was the method of expressing concentration, but when "per cent solute" was used the value for specific gravity was multiplied by 10^3 and the indicated percentage of this value yielded the concentration in grams per liter. The value for concentration in terms of grams per liter was divided by the respective fraction $\frac{Wa}{Wh}$, derived as the sums of the values derived for the involved ions through the use of the description of hydration. This procedure yielded values which represented the weight of the hydrated solute in terms of grams per liter. The weight of the hydrated solute was then divided by D, or $1 + \frac{Wa}{Wh}$, to yield a value representing the volume of the hydrated solute in milliliters. The value thus obtained was subtracted from 10^3 to yield the volume of the solvent in milliliters, a value which also was the weight of the solvent in grams. This value then was added to the weight of the hydrated solute and the resulting sum was divided by 10^3 to yield the calculated or predicted specific gravity value of the solution.

In conjunction with the foregoing it was obvious that an agreement between calculated and observed specific gravity values was conditioned not only by the integrity of the assumptions relating to the solute ions present in the solution but also by the integrity of the appraisals of their respective specific hydrational potentialities. As thus represented the specific gravity of an aqueous solution, became an index of the nature and state of the involved solute units.

The interesting, important and dramatic feature of this development became evident when calculated or predicted specific gravity values were compared with observed values. Here was a wealth of observational data readily available to anyone, and the solute concentrations cited did not involve atomic weight values. On this account anyone of a mind to substitute conventional atomic weight values, irregular and fractional, for the integral atomic weight values prescribed by the description of periodic hydrational potentiality certainly was free to do so. In fact, he was urged and challenged to do so: there was envisioned no simpler procedure for convincing the skeptical conventional chemist of the integrity of the basic order inherent in the description of periodic hydrational potentiality.

Data affording comparisons of calculated and observed specific gravity values for specified aqueous solutions have been assembled to comprise Tables 1 and 2. In the column headings for these tables Z=summation, a=anhydrous ions, h=completely hydrated ions, W =weight.

Collectively the data of these two tables and the indicated excellent order of agreement between calculated and observed specific gravity values sustained both the integrity of the investigators responsible for the data and the integrity of the involved assumptions. There was no alternative to the conclusion that in aqueous solutions specific gravity was an index of hydration, to the conclusion that all of the assumed ions had been present in the respective solutions, to the conclusion that the ions had evidenced the weights and hydrational potentialities prescribed for them by the description of periodic hydration. The tables were interpreted as having documented the validity of the description of periodic hydrational potentiality.

Table 1

Data affording a comparison of observed specific gravity values with values calculated with the description of periodic hydrational potentiality. Solutes as element ions.

Solute	W_a		Z_a	W_h		Z_h	$\frac{Z_a}{Z_h}$	Concentration gms./Liter	Calculated Specific Gravity	Observed Specific Gravity
AlCl₃	Al+3 = 32 Cl- = 32		128	Al+3 = 158 Cl- = 158		632	.2025	85.69	1.071	1.0711
CuCl₂	Cu++ = 62 Cl- = 32		126	Cu+2 = 332 Cl- = 158		648	.195	241.0	1.204	1.205
ZnCl₂	Zn++ = 64 Cl- = 32		128	Zn++ = 316 Cl- = 158		632	.203	41.4	1.0348	1.035
SnCl₂	Sn+2 = 104 Cl- = 32		168	Sn++ = 410 Cl- = 158		726	.231	131.8	1.107	1.0986
SnCl₄	Sn+4 = 108 Cl- = 32		236	Sn+4 = 378 Cl- = 158		1010	.2236	131.9	1.106	1.099
CsCl	Cs+ = 112 Cl- = 32		144	Cs+ = 346 Cl- = 158		504	.2857	169.4	1.1222	1.1217
BaCl₂	Ba++ = 116 Cl- = 32		180	Ba++ = 314 Cl- = 158		630	.2857	240.6	1.191	1.2031
AuCl₃	Au+ = 164 Cl- = 32		260	Au+3 = 344 Cl- = 158		818	.3175	107.5	1.081	1.075
HgCl	Hg+2 = 164 Cl- = 32		228	Hg++ = 344 Cl- = 158		660	.3454	52.055	1.0403	1.0411

Table 2

Data affording a comparison of observed specific gravity values with values calculated with the description of periodic hydration. Solutes as element ions.

Solute	W_a		Z_a	W_h		Z_h	$\dfrac{Z_a}{Z_h}$	Concentration gm./Liter	Calculated Specific Gravity	Observed Specific Gravity
Na_2S	Na^+ = 24		76	Na^+	= 222	634	.1205	136.7	1.125	1.128
	S^{--} = 28			S^{--}	= 190					
$MgBr_2$	Mg^{++} = 28		164	Mg^{++}	= 190	758	.2163	184.256	1.1515	1.1516
	Br^- = 68			Br^-	= 284					
KI	K^+ = 40		144	K^+	= 94	504	.285	206.4	1.153	1.1468
	I^- =104			I^-	= 410					
AgF	Ag^+ = 96		112	Ag^+	= 474	760	.1473	135.792	1.1168	1.1316
	F^- = 16			F^-	= 286					
CdI_2	Cd^{++} =100		308	Cd^{++}	= 442	1262	.2440	315.2	1.2515	1.2608
	I^- =104			I^-	= 410					
$MnBr_2$	Mn^{++} = 54		190	Mn^{++}	= 396	964	.197	238.84	1.200	1.194
	Br^- = 68			Br^-	= 284					
CsI	Cs^+ =112		216	Cs^+	= 346	756	.2857	389.83	1.303	1.2994
	I^- =104			I^-	= 410					
WO_3	W^{+6} =160		195	W^{+6}	= 376	1330	.1473	243.4	1.2111	1.217
	O^{--} = 12			O^{--}	= 318					

The solutes selected for inclusion in the foregoing tables were assumed to have yielded element ions in aqueous solution but the description of periodic hydrational potentiality related the attractive force to ionic weight. From the viewpoint of a substantiation of the Prout hypothesis the inclusion of the item (A.N.+C.) in the description of hydration was of primary importance, but if ionic weight was the critical attribute conditioning hydration the hydrational potentiality of solute radicals was subject to projection as precisely that which, under the description of hydration, would characterize an element ion of the same weight. In the event that the specific gravity of aqueous solutions was the reliable index of hydration which had been interpreted as evidenced for solute element ions it was of potential service in indicating the nature and state of solute radicals. Data for some solutes assumed to yield both element ions and radicals in aqueous solution have been given in Tables 3 and 4.

An examination of the data of Tables 3 and 4 made it clear that the periodic nature of hydrational potentiality, in contrast to the periodicity of chemical properties represented in the Mendeleev system and associated with orbital status, was a function of ionic weight. This was considered to be an important development because it seemed to indicate that the involved attractive force, although instituted and limited by ionization, was mediated fundamentally by the atomic weight.

Included in the four preceding tables were data for element ions and radicals of weight values representing each of the four regular hydration periods. It became of interest to examine these data in relation to the four periods, each of which involved an inverse relationship between ionic weight and hydrational potentiality. The results obtained have been indicated in Table 5.

The data given in Table 5 indicated the distribution of the solute units which had been given in the preceding four tables within the respective four periods of the hydration system described in chapter 2. The data also indicated that within each period there was a seriate inverse relationship between the weights in the anhydrous state, the Wa values, and the numbers of H_2O^- units potentially taken on in hydration, the H values. It was obvious that the involved solutes had included solute units representative of each of the four periods. Had it seemed desirable, the list could have been extended, and both calculated and observed specific gravity values for all solutes at an extensive series of concentrations could have been cited.

For a general appraisal of the data given in the five tables, it was held that there was no reason to question the integrity of the observed specific gravity values, since these had been obtained through the researches of well-trained and meticulous scientists. It was held further that no successful calculation of specific gravity values for aqueous solutions of any of the listed solutes was possible independent of the description of periodic hydration. It was held that in view of the fact that different investigators and different temperatures were represented in the observational data the degree of agreement between calculated and observed specific gravity values was far beyond that which could be attributed to chance. It was concluded, therefore, that the specific gravity of aqueous solutions forthwith had documented the integrity of the description of periodic hydration. Inseparable from this conclusion was the evidence that the specific gravity of aqueous solutions by virtue of the description of periodic hydrational potentiality had attained recognition as a reliable index of the nature and status of the involved solute units.

Within the scope of this chapter the validation of the description of periodic hydration was held to be the most important service that specific gravity as an index could render. For that reason the solutes of the tabulated data were restricted to those evidenced by specific gravity as yielding hydrated solute units exclusively in aqueous solution. It was noted, however, that specific gravity as an index of solute nature and status was of service in relation to other matters. One of these concerned the attribute of potential acid-alkali reactivity. Another concerned hydrational bondage in the absence of free solvent. These matters and other allied subjects will be taken up in subsequent chapters.

Ordinarily the documentation of the integrity of the description of periodic hydration as represented by the data of Tables 1-5 might be expected to prove convincing and acceptable, especially since the observational data had been obtained by highly reputable investigators. The description of periodic hydration, however, held serious implications with respect to the integrity of long-esteemed atomic weight values. In the data of Table 2, for example, the evidenced ionic weight values $O^{--}=12$ and $W^{+6}=160$ would scarcely be expected to meet with universal favor among contemporary chemists. There was no disposition to rest the integrity of the description of periodic hydration solely upon the evidence afforded by the specific gravity of aqueous solutions.

Table 3

Data affording a comparison of observed specific gravity values with values calculated with the description of periodic hydrational potentiality. Solutes include radicals.

Solute	W_a	W_b	Z_a	Z_b	$\frac{Z_a}{Z_b}$	Concentration gms/litter	Calculated Specific Gravity	Observed Specific Gravity
$MgSO_4$	$Mg^{++} = 28$ $SO_4^{--} = 92$	$Mg^{++} = 190$ $SO_4^{--} = 506$	120	696	.172	337.0	1.290	1.2961
K_2SO_4	$K^+ = 40$ $SO_4^{--} = 92$	$K^+ = 94$ $SO_4^{--} = 506$	172	692	.248	108.17	1.086	1.0817
Na_2SO_4	$Na^+ = 24$ $SO_4^{--} = 92$	$Na^+ = 222$ $SO_4^{--} = 506$	140	950	.1472	85.79	1.074	1.0724
Na_2SO_3	$Na^+ = 24$ $SO_3^{--} = 76$	$Na^+ = 222$ $SO_3^{--} = 220$	124	654	.1859	211.6	1.177	1.1755
$FeSO_4$	$Fe^{++} = 56$ $SO_4^{--} = 92$	$Fe^{++} = 380$ $SO_4^{--} = 506$	148	885	.167	242.7	1.205	1.210
$Fe_2(SO_4)_3$	$Fe^{+3} = 58$ $SO_4^{--} = 92$	$Fe^{+3} = 364$ $SO_4^{--} = 506$	392	2246	.1745	85.36	1.073	1.067
$MnSO_4$	$Mn^{++} = 54$ $SO_4^{--} = 92$	$Mn^{++} = 396$ $SO_4^{--} = 506$	146	902	.162	215.1	1.185	1.195

Table 4

Data affording a comparison of observed specific gravity values with values calculated with the description of periodic hydrational potentiality. Solutes include radicals.

Solute	W_a	Z_a	W_b	Z_b	$\dfrac{Z_a}{Z_b}$	Concentration gms/liter	Calculated Specific Gravity	Observed Specific Gravity
K_2CO_3	$K^+ = 40$ $CO_3^{--} = 56$	136	$K^+ = 94$ $CO^{--}_3 = 380$	568	.2394	159.026	1.1273	1.1359
$Ni(NO_3)_2$	$Ni^+ = 60$ $NO_3 = 60$	180	$Ni^+ = 348$ $NO^-_3 = 348$	1044	.1720	41.32	1.035	1.033
$Sr(NO_3)_2$	$Sr^{++} = 80$ $NO_3 = 60$	200	$Sr^{++} = 188$ $NO^-_3 = 348$	884	.2262	237.40	1.1949	1.1870
NaH_2ASO_4	$Na^+ = 24$ $H_2ASO^-_4 = 132$	156	$Na^+ = 222$ $H_2ASO^-_4 = 186$	408	.3823	156.142	1.1127	1.1153
Na_2MoO_4	$Na^+ = 24$ $MoO^{--}_4 = 144$	192	$Na^+ = 222$ $MoO^{--}_4 = 504$	948	.2025	388.60	1.1995	1.1943
Li_2IO_3	$Li^+ = 8$ $IO^-_3 = 152$	160	$Li^+ = 350$ $IO^-_3 = 440$	790	.2025	184.88	1.1535	1.1555
$Pb(NO_3)_2$	$Pb^{++} = 168$ $NO^-_3 = 60$	288	$Pb^{++} = 312$ $NO^-_3 = 348$	1008	.2865	41.375	1.032	1.0344
$K_2S_2O_7$	$K^+ = 40$ $S_2O_7^{--} = 172$	252	$K^+ = 94$ $S_2O_7^{--} = 280$	468	.5384	258.698	1.1669	1.1759

Table 5

Data for the solute ions included in the preceding tables, arranged to indicate the relation of their evidenced hydration to the description of periodic hydrational potentiality.

First Hydration Period			Second Hydration Period			Third Hydration Period			Fourth Hydration Period		
Ions	W_a	H	Ions	W_a	H	Ions	W_a	H	Ions	W_a	H
Li^+	8	19	Mn^{++}	54	19	SO_4^{--}	92	23	MoO_4^{--}	144	20
O^{--}	12	17	Fe^{+2}, CO_3^{--}	56	18	Ag^+	95	21	IO_3^-	152	16
F^-	16	15	Fe^{+3}	58	17	Cd^{++}	100	19	W^{+6}	160	12
Na^+	24	11	Ni^{+2}, NO_3^-	60	16	$Sn^{++}, 1-$	104	17	Au^{+3}, Hg_9^{+2}	164	10
Mg^{++}, S^{--}	28	9	Cu^{++}	62	15	Sn^{+4}	108	15	Pb^{+2}	168	8
Al^{+3}, Cl^-	32	7	Zn^{++}	64	14	Cs^+	112	13	$S_2O_3^{--}$	172	6
K^+	40	3	Br^-	68	12	Ba^{++}	116	11			
			SO_3^{--}	76	8	$H_2ASO_4^-$	132	3			
			SM^{++}	80	6						

CHAPTER 4
OSMOSIS

As one of the best known and least understood processes involving water osmosis long engendered a controversial salient in research from a diffuse front in physics, chemistry and physiology. For reasons which will become apparent the process had to remain enigmatic in the absence of a recognition of hydrational potentiality as a periodic function of ionic weight.

At the outset one well may appraise and applaud the classical researches of PFEFFER during the golden era of German science in the closing quarter of the preceding century. The work of PFEFFER not only emphasized the importance of osmosis as a process intimately involved in physiology but also established a pattern of inquiry which dominated research in the area for many years. This pattern stressed the dramatic feature of osmosis and for the most part even present day osmometers in design and function still reflect the early emphasis on pressure. The word "osmosis" itself was coined from a Greek word meaning "a pushing", and throughout the years the process has been appraised within the framework of the original context.

Throughout the years, moreover, there has been a recognition of the potential importance of a more adequate knowledge of osmosis. Inherent in every living cell, plant or animal, there was envisioned the ability to form a vacuole potentially active in osmosis: the osmometers used in laboratory research have been copies of the liquid-filled vacuoles in nature. The diversity of form and physiology among plants and animals may be appraised as practically infinite, and vacuoles as adjuncts to protoplasmic metabolism certainly might be expected to share in this diversity. For the present, however, any such diversity need be of no concern, for in the laboratory with a uniform osmotic membrane the behavior patterns of different solutes have been indicated as sufficiently different, interesting and challenging to engage research energies for some time to come. In years past the challenge of these behavior patterns has led to a compounding of confusion which testified only to the resourcefulness of frustrated investigators. It would be interesting but digressive, distracting and time-consuming to include here a review of the literature on osmosis since the time of PFEFFER. The integrity of any description in science long has rested upon its serviceability in prediction and heretofore, although there have been differences in approach involving temperature, vapor pressure, nature of solute, concentration of solute, properties of the interposed membranes and various types of factor inter-relationships no description of osmosis has supplied a satisfactory basis for the successful prediction of osmotic behavior patterns.

By what might seem an usual fortuitousness the description of periodic hydrational potentiality given in Chapter 2 when integrated with specific observations presently to be considered yielded a description of osmosis quite different from any preceding description and repeatedly demonstrable as exceedingly efficient in prediction. The new description of osmosis evolved through the adoption of a practice long used in the study of the behavior of gases: the establishment of standardization conditions. Quite arbitrarily for convenience these standard conditions were: the increase in solution volume of one liter of a 1.0 molar solution at 25°C and 760 mm atmospheric pressure. As thus established ionic atomic or molecular weight values were transposable to grams, ionic volume values were transposable to milliliters and for the aqueous solvent grams and milliliters were interchangeable. It was true that no osmometer used had a capacity of 1 liter; and some solutes were not soluble to an extent sufficient to provide a solution of 1.0 molar concentration. These latter considerations, however, did not interfere with the practical advantages insured by the standardization of the observational data. The new description evolved as a result of researches on osmosis which extended over a period of several years. There is no implication that either the description or its allied concepts are infallible or universally applicable or incapable of improvement.

The osmometers used exclusively in the early researches and extensively at all times were obtained from the Central Scientific Co., Chicago, Ill., and were designated as of the double-faced Cenco-Troxel type. These were of metal and had a capacity of somewhat less than 100 ml. They appeared to have been designed primarily for demonstration purposes rather than research, but served admirably insofar as the evolution of a satisfactory description of osmosis was concerned. In the initial research an absorptive cellophane was used as the membrane. Subsequently a firmer cellulose casing was used and preferred as being the more durable in successive experiments. The absorptive membranes came to be appraised as semi-permeable only in a specially-restricted sense; permeable to anhydrous ions but not to hydrated ions. The appraisal involved the concept of water as of a greater molecular complexity than that represented by the symbol H_2O, but as yielding to the absorptive membrane matrix the negatively charged H_2O^- units involved in the hydration of solutes. On the part of a solution the solute ions irrespective of their status with respect to hydration were interpreted as capable of entering the absorptive matrix only in the anhydrous state. As relatively more mobile ions within the absorptive matrix they became hydrated therein through contact with the H_2O^- units. These solute ions when completely hydrated forthwith made a randomized departure at all membrane-liquid interfaces. As thus represented the process continued until one-half the solute, in a system originally with a solution on the inside and water on the outside, had returned to the interior of the osmometer and one-half the solute had been extruded to the bathing liquid. The increase in solution volume, therefore, was occasioned by the return of one-half the original solute in the hydrated state and the culmination of this process represented the maximum increase in solution volume and the termination of osmotic activity. At this termination in osmosis as ordinarily involved in the laboratory the extruded half of the hydrated solute would be dispersed in a larger volume of bathing liquid than the hydrated solute within the osmometer. Under such circumstances osmosis was evidenced as having been followed by a process of adjustment involving a departure of ions in the anhydrous state from the internal liquid to the external or bathing liquid. This movement continued until the solute concentration was uniform throughout the internal and external liquids and resulted in slight reductions in the volume of the internal liquid. At the conclusion of this process the internal solution volume remained constant, irrespective of the height of the liquid above that of the bathing liquid. The experiments relating to this constancy were discontinued after thirty months, at which time it was concluded that the state was real or apparent equilibrium would continue indefinitely. In tests so designed as to insure equal volumes of internal and external liquids at the termination of osmosis there was no decrease in the volume of the internal liquid and the maximum increase in internal volume attributable to osmosis was sustained as the increase characterizing the stabilized state. It was recognized that special osmometers could be devised to facilitate the measurement of osmosis exclusively, but with both the indicated osmometers and similar all-plastic osmometers designed to circumvent ionic reactions with metal the corrections for supplementary ionic movement when specific external and internal liquid volumes were involved presented no difficulty.

In conjunction with the foregoing representation of osmotic activity it was obvious that the aqueous solvent possessed an attraction for the H_2O^- units involved in the hydration of the solute ions and that the absorptive matrix of the membrane possessed a greater attraction for these units. For a long time it had been recognized that water possessed potentialities for dissociating some chemical compounds. In a later chapter it will be made apparent that the absorptive membrane matrix possesses greater potentialities than water for dissociating some chemical compounds. Of more immediate concern at this point, however, is the fact that in all of the writer's researches on osmosis the evidenced general osmotic behavior pattern was the same: at the time of the attainment of maximum solution volumes within the osmometers equal amounts of solute were in the internal and external liquids. The importance of this exudative aspect of osmosis in relation to physiology will become quite apparent without special emphasis at this time.

With regard to a more specific appraisal of osmosis it appeared that there was no direct passage of water as such from the external aqueous liquid into the osmometers: water entered only by virtue of its association with ionic solutes within the absorptive membrane and the randomized exit from the membrane which returned one half of the original solute back into the osmometers. On the other hand, following the completion of osmotic activity, whenever the volume of the external liquid was greater than the volume of the internal liquid there took place, as noted, a redistribution to effect equal concentrations of solute in the internal and external liquids. Incident to osmosis, therefore, there was a passage of hydrationally-bonded water into the osmometers and with common usage there was a release internally of some of the water incident to the establishment of uniform solute concentrations. As thus appraised neither osmosis nor the supplementary redistribution of solute involved a direct passage of water through the membrane.

The foregoing interpretations were derived from studies of behavior patterns involving solutions in osmometers immersed in water. Quite naturally it became of interest to study the behavior patterns when different solutions were present initially on opposite sides of an absorptive membrane. Such studies were carried out through the use of opposing solutions assumed to supply ions which would from visible precipitates on contact, as solutions of $AgNO_3$ and $NaCl$, $Ba(NO_3)_2$ and $MgSO_4$, $Ca(NO_3)_2$ and K_3PO_4. Under such conditions precipitates could be demonstrated readily in both internal and external solutions. It was concluded, therefore, that the solutes had entered the absorptive membrane from both sides. The volumetric changes which were associated with osmotic activity involving precipitation were such as to evidence a release of the water of hydration incident to precipitation and the consequent characterization of the precipitate as anhydrous. It seemed obvious that hydration did not interfere with chemical reactivity. The relation of these developments to the absorptive uptake of ions and to physiology will be discussed in a subsequent chapter.

The foregoing considerations may be restated in terms of relative attraction. An absorptive membrane was evidenced as having a greater attraction for H_2O^- sub-units of water than the attraction exerted by liquid water itself and hence as capable of withdrawing the units from water. The solute ions were evidenced as having a greater attraction for the H_2O^- units than the absorptive membrane and hence as capable of withdrawing these units from the matrix. The extent of the attraction exerted by the solute ions was evidenced as conditioned by ionic weight and the extrusion of hydrated ions from the matrix appeared to mediate osmosis. As thus appraised it was obvious that osmosis had potentialities for refuting or validating the integrity of the description of periodic hydration. Moreover, as thus appraised it was obvious that an osmotic validation of the description would afford an explanation of the consummate difficulties which had been encountered in the numerous attempts of other investigators to obtain a satisfactory interpretation of osmosis. The relative importance of osmotic behavior patterns thus seemed to depend upon the point of view. In the event that the observed specific gravity of aqueous solutions as considered in the preceding chapter was appraised as having adequately documented the validity of the description of hydration the importance of the description was the potential service in the interpretation of osmosis. Alternatively, the importance of osmosis was its potential service in supplementary validation of the description of hydration. To the writer the important thing was the evidenced role of hydration in osmosis, on which account it was understandable that in the absence of an adequate knowledge of hydration it had been impossible to derive any satisfactory description of osmotic behavior patterns.

The description of periodic hydrational potentiality as given in the second chapter was oriented to the electrometric data of KOHLRAUSCH involving mobilities of solute ions under stress. The correlation of these data with the Graham law of diffusion and the ABEGG and BODLANDER principle of electro-affinity involved the concept of the attractive force mediating hydration as being sufficient to allow the solute ions to retain the attracted H_2O^- units against the stresses imposed incident to the taking of the electrical measurements. The mobility values thus became subject to correlation with the weights of the solute ions in hydrated states.

It was assumed that in the measurement of the specific gravity of aqueous solutions the stresses imposed would be less than those involved in the electrical measurements, but there had remained the possibility that the electrical tension might have induced hydration or modified it. There was substantial satisfaction, therefore, when it was found, as evidenced in the data of Chapter 3, that the solute ions in the listed solutions were present in hydrated states which had not been brought about by stresses other than those involved in effecting solution.

The fact that aqueous solutions had been evidenced as containing hydrated ions seemed to hold promise in conjunction with an analysis and interpretation of the behavior patterns of solute and solvent in osmosis. These behavior patterns presented some aspects out of accord with conventional view points regarding osmosis, but since these conventional viewpoints were diverse and controversial there was little concern over the situation. The important project seemed to be to determine whether or not the validated description of periodic hydrational potentiality could contribute to an improved understanding of osmosis. In attempting to solve this problem a somewhat surprising development took place. Not only did the description of periodic hydrational potentiality afford an improved understanding of osmosis: it also revealed that osmosis was of substantial service in supplying a further means of documenting the integrity of the description.

In the projected correlation of hydration and osmosis the relation of volume in the anhydrous state to volume in the hydrated state again assumed importance. The successful solution of this problem had preceded the initiation of researches on osmosis, and without such a solution there could not have evolved any definitive recognition of hydration as having a dominant role in osmosis. It was the successful solution of this problem which made possible the recognition of specific gravity as an index of ionic weight and hydration in aqueous solutions, as was made evident in the preceding chapter.

The density of hydrated solute ions relative to an aqueous solvent was found and reported to be subject to description as

$$D = 1.0 + \frac{Wa}{Wh}$$

When D=density, Wa=weight of anhydrous ion, and Wh=weight of hydrated ion. This relationship obviously was a simple one; but as obviously it was not a description which could be derived prior to an adequate description of hydration. Less obviously the relationship was not subject to implementation apart from adequate procedures for the appraisal of the volumetric changes accompanying transitions from anhydrous to hydrated states or the reverse.

By extraordinary good fortune the standardized observational data of osmosis as interpreted in conjunction with the description of periodic hydrational potentiality and the description of solute density brought to the study of the behavior patterns of solute ions a mathematical background analogous to that represented in the gas laws. As indicated previously, it was the latter background which had prompted PROUT to advance his hypothesis regarding the interrelationship and common origin of the elements. Of immediate practical importance was the fact that the mathematical background made available for the study of the behavior patterns of solute ions through the integration of the attributes of weight, volume and mobility had implemented the description of periodic hydrational potentiality and its corollaries with the most trustworthy criterion of integrity known to science: facility in prediction.

In keeping with the foregoing considerations some data for sodium chloride appraised as a representative solute have been presented. Observational data for the osmotic behavior of the solute have been given in Table 6.

In the data of Table 6 it was to be noted that the standardized osmotic increase in volume was about the same at all the listed concentrations. This meant that each aliquot increment of solute contributed equally to the volumetric increase. PFEFFER and a host of other investigators had obtained analogous results. To an appreciable extent such results had engendered the concept of independence for the solute units and a consequent analogy with gases. Without an adequate description of hydration there was no prospect for further interpretation.

The situation was quite different, however, following the development of the description of periodic hydrational potentiality given in Chapter 2. This description was integrated with the previously-indicated behavior patterns in such a manner as to provide a procedure for calculating or predicting osmotic potentialities. Such a procedure has been given in Table 7 for sodium chloride.

Table 6

Results obtained in a study of the osmotic behavior of sodium chloride.
Temperature 25°C, Mol. Wt. 56. Initial Solution Volumes, 75 ml.

NaCl grams salt	NaCl Molar Conc as prepared	Observed Volume in Milliliters at completion of Osmosis	Calculated Volumetric Increase in milliliters	Calculated Standardized Osmotic Increase in Volume
2.1	.5	80	5.0	133.3
4.2	1.0	85.5	10.5	140.0
6.3	1.5	90.5	15.5	137.7
8.4	2.0	95.5	20.5	136.5
10.5	2.5	101	26.0	138.6
12.6	3.0	106	31.0	137.7
16.8	4.0	116	41.0	136.7
21.0	5.0	126.5	51.5	137.3

Table 7

Data indicating the procedures followed in calculating or predicting
the osmotic behavior pattern for sodium chloride.

Assumed Ions	Wa	n	H	Wh	Vh	Va	D_1	$\dfrac{D_1}{2}$	$\dfrac{Va}{2}$	D_2	u	V_1	Calculated Standard Increase
Na+	24	1	11	222	200	12	188	94	6	88	1	88	
Cl-	32	1	7	158	131	16	115	57.5	8	49.5	1	49.5	137.5

Supplementary Legend : Vh=volume of hydrated ion; Va=volume of anhydrous ion;
D_1=difference, Vh—Va; u=number of ionic units per molecule; V_1=volume per ion.

The procedures indicated in Table 7 appeared to merit discussion in some detail because the pattern will be followed in subsequent considerations of osmotic behavior. The values in the column headed Wa were derived by use of the second corollary of the description of periodic hydration. The values of the "n" and "H" columns were derived directly from the description. The values of the "Wh" column were derived by the use of the fourth corollary of the description of periodic hydration. The values of the "Vh" column were derived from those of the "Wh" column through the use of the formula for density given previously in this chapter. The values of the "Va" column were derived by dividing the Wa values by 2, the uniform density of anhydrous ions indicated by the density formula. The values of the "D_1" column were derived as Vh—Va and represent the differences in volume in the anhydrous and hydrated states. The values of column "D_2" were derived as one-half of the respective "D_1" column values and relate to the volumetric increase effected by the return of half the ions to the osmometer in the hydrated state. The values of colomn Va/2 were derived as one-half the values of column "Va" and relate to the decrease in volume associated with the permanent loss of one-half the number of ions originally present with the osmometer. The values of the column headed "D_2" were derived as the differences in the values of the two preceding columns and represent the net internal volumetric increase attributable to the involved specified ion. The values of the column headed "u" relate to the number of units of the specified ion in a compound. It would have importance, for example in such a compound as magnesium chloride, $MgCl_2$. The values of the column headed "V_1" were derived as the products of the values of the two preceding columns. The value of the column headed "V_T" was derived as the sum of the values in the preceding column and represented the calculated or predicted total volumetric increase inside the osmometor attributable to osmosis. In Table 7 this value was subject to comparison with the observed standardized increase in volume given in the right hand column of Table 6.

For any analyst particularly interested in statistical evaluations it might be stated that with respect to the data of Table 6 and 7 the degree of correlation between calculated and observed values could be enhanced by making appropriate modifications of the procedures followed in obtaining the observational data. It seemed preferable, however, to use uniform procedures.

With respect to the use of 56 as the molecular weight of sodium chloride it was to be noted that the use of a contrasting conventional molecular weight of the order of 58.454 in Table 6 would have yielded standardized osmotic increases of about 143.5 instead of about 137.5. The use of such an ionic weight value as 36 instead of 32 for the Cl⁻ ion in Table 7 would have yielded a calculated standardized osmotic increase of about 120 instead of about 137.5. Agreements were to be obtained, therefore, only when weight values prescribed by the description of periodic hydrational potentiality were used. The fundamental and important significance of this situation became increasingly obvious when analogous data for other solutes were obtained and examined. Such an examination will comprise the context of the chapter which follows.

For the characterization of ions as involved in the data of Table 7 it was obvious that a great deal of repetition would be eliminated if the major attributes of all ions over the O—184 weight range was recorded in tabular form. With the exception of volume and osmotic behavior such attributes were published previously [see references at end of this chapter]. A more complete tabulation is included herewith in order to facilitate the calculation of osmotic behavior patterns. In the following Table 8 the data have been separated into four groups, each group representing one period of hydrational potentiality.

The data given in Table 8 were compiled to serve as an aid in calculating or predicting behavior patterns for atomic ions, but since hydration was evidenced as conditioned by ionic weight exclusively the data proved equally helpful in studies involving solute radicals and ionic amphoteric non-electrolytic solutes. The legend for the column headings is as follows, left to right: A.N.=atomic number, E=element, Wa=weight as a neutral atom, anhydrous, to the base O_2=32, Ma=mobility under the same conditions, H = number of H_2O^- units potentially taken on and held by ions of the indicated Wa value, Wh=weight of ions in the hydrated state, Vh =volume of hydrated ions, Mh=mobility of hydrated ions and Vi=volumetric increase attributable to the osmotic behavior of the ion under standard conditions, —in milliliters.

With specific reference to osmosis the convenience afforded by the data of Table 8 can best be illustrated by example. In the event that the osmotic behavior of sodium iodide, NaI, was to be calculated, the procedure would be as follows. The assumption would be made that in aqueous solution the ions present would be Na^+ and I^-. From Table 8A in conjunction with the description of

Table 8 A.   Reference data for the first hydration period.

A.N.	E	Wa	Ma	H	Wh	Vh	Mh	Vi
0				23	414	414	491	207
1	H	2	7082	22	398	379	501	188.5
2	He	4	5000	21	382	365	512	180.5
3	Li	6	4090	20	366	360	523	177
4	Be	8	3546	19	350	342	535	167
5	B	10	3162	18	334	325	547	157.5
6	C	12	2887	17	318	306	561	147
7	N	14	2672	16	302	289	575	137.5
8	O	16	2500	15	286	271	591	127.5
9	F	18	2357	14	270	253	609	117.5
10	Ne	20	2236	13	254	236	627	108
11	Na	22	2132	12	238	218	648	98
12	Mg	24	2041	11	222	200	671	88
13	Al	26	1961	10	206	183	698	78.5
14	Si	28	1890	9	190	166	726	69
15	P	30	1826	8	174	148	758	59
16	S	32	1768	7	158	131	796	49.5
17	Cl	34	1715	6	142	115	839	40.5
18	A	36	1667	5	126	98	891	31
19	K	38	1622	4	110	82	953	22
20	Ca	40	1581	3	94	66	1031	13
21	Sc	42	1543	2	78	51	1133	4.5
22	Ti	44	1508	1	62	36	1271	−4
23	V	46	1474	0	46			

Table 8 B.   Reference data for the second hydration period.

A.N.	E	Wa	Ma	H	Wh	Vh	Mh	Vi
23	V	46	1475	23	460	420	466	187
24	Cr	48	1443	22	444	410	475	181
25	Mn	50	1414	21	428	383	484	166.5
26	Fe	52	1387	20	412	361	494	153.5
27	Co	54	1360	19	396	348	502	147
28	Ni	56	1336	18	380	331	513	136.5
29	Cu	58	1313	17	364	314	524	128
30	Zn	60	1290	16	348	297	536	118.5
31	Ga	62	1270	15	332	280	549	109
32	Ge	64	1250	14	316	263	563	99.5
33	As	66	1230	13	300	246	578	90
34	Se	68	1212	12	284	229	594	80.5
35	Br	70	1195	11	268	212	611	71
36	Kr	72	1178	10	252	196	630	62
37	Rb	74	1162	9	236	180	651	53
38	Sr	76	1147	8	220	163	675	43.5
39	Y	78	1132	7	204	147	700	36.5
40	Zr	80	1118	6	188	132	729	26
41	Cb	82	1105	5	172	116	763	12
42	Mo	84	1091	4	156	101	800	8.5
43	Ma	86	1078	3	140	87	846	0.5
44	Ru	88	1065	2	124	72	899	−8
45	Rh	90	1054	1	108	59	963	−15.5
46	Pd	92	1042	0	92			

Table 8 C. Reference data for the third hydration period.

A.N.	E	Wa	Ma	H	Wh	Vh	Mh	Vi
46	Pd	92	1042	23	506	430	445	169
47	Ag	94	1031	22	490	410	452	158
48	Cd	96	1020	21	474	394	460	149
49	In	98	1010	20	458	377	467	139.5
50	Sn	100	1000	19	442	361	476	130.5
51	Sb	102	992	18	426	344	485	121
52	Te	104	982	17	410	327	494	111.5
53	I	106	972	16	394	310	502	107
54	Xe	108	963	15	378	294	514	93
55	Cs	110	954	14	362	278	526	84
56	Ba	112	945	13	346	261	538	74.5
57	La	114	936	12	330	245	551	65.5
58	Ce	116	929	11	314	229	565	56.5
59	Pr	118	922	10	298	219	580	50.5
60	Nd	120	914	9	282	198	595	39
61	Il	122	906	8	266	182	613	30
62	Sm	124	898	7	250	167	631	21.5
63	Eu	126	891	6	234	152	654	13
64	Gd	128	884	5	218	137	678	4.5
65	Tb	130	878	4	202	123	704	−3.5
66	Dy	132	871	3	186	109	733	−11.5
67	Ho	134	864	2	170	95	768	−19.5
68	Er	136	858	1	154	82	806	−27
69	Tm	138	851	0	138			

Table 8 D. Reference data for the fourth hydration period.

A.N.	E	Wa	Ma	H	Wh	Vh	Mh	Vi
69	Tm	138	851	23	552	451	426	151.5
70	Yb	140	845	22	536	425	432	142.5
71	Lu	142	839	21	520	408	439	133
72	Hf	144	833	20	504	392	445	124
73	Ta	146	828	19	488	376	453	115
74	W	148	822	18	472	359	460	105.5
75	Re	150	817	17	456	343	468	96.5
76	Os	152	811	16	440	327	477	87.5
77	Ir	154	806	15	424	311	486	78.5
78	Pt	156	801	14	408	295	495	69.5
79	Au	158	796	13	392	279	505	60.5
80	Hg	160	790	12	376	264	516	52
81	Tl	162	786	11	360	246	527	42
82	Pb	164	781	10	344	233	539	34.5
83	Bi	166	776	9	328	218	552	26
84	Po	168	772	8	312	203	566	17.5
85		170	768	7	296	188	581	9
86	Rn	172	763	6	280	173	598	0.5
87		174	758	5	264	159	615	−7.5
88	Ra	176	754	4	248	145	635	−15.5
89	Ac	178	750	3	232	131	657	−23.5
90	Th	180	746	2	216	118	680	−31
91	Pa	182	741	1	200	105	707	−38.5
92	U	184	737	0				

periodic hydrational potentiality the weight of the Na^+ ion would be subject to derivation as 24 and the volumetric increase attributable to Na^+ ions under standard conditions would be indicated in the right hand column as 88 ml. From Table 8 C in like manner the weight of the I^- ion would be subject to derivation as 104 and the volumetric increase attributable to I^- ions under standard conditions would be indicated as 111.5 ml. The calculated osmotic volumetric increase for NaI would be the sum of the indicated values for the Na^+ and I^- ions, or 199.5 ml.

Several subsequent chapters of this book will involve considerations of osmotic behavior, and since the procedures initially will be unfamiliar the steps involved have been given in detail in the tabulations. It is to be recognized, however, that with the data of Table 8 at hand the calculation or prediction of osmotic behavior patterns in conjunction with the description of periodic hydrational potentiality becomes a relatively simple procedure whose success or failure is conditioned basically by the correctness of the assumptions made regarding the solute ions involved.

LITERATURE CITED

FLINT, L. H. (1932). Hydration of the solute ions of the lighter elements. Jour. Wash. Acad. Sci. 22:97-119.

—(1932). Hydration of the solute ions of the heavier elements. Jour. Wash. Acad. Sci. 22:211-217.

PFEFFER, W. (1877). Osmotische Untersuchungen. W. Englemann. Leipzig.

CHAPTER 5

OSMOSIS AS AN INDEX OF PERIODIC HYDRATIONAL POTENTIALITY

The analysis of the observational data on the specific gravity of aqueous solutions as given in Chapter 3 was interpreted as having documented the validity of the description of periodic hydrational potentiality. An extension of the analysis revealed that at higher concentrations of many solutes the observational data evidenced that a partial or complete satisfaction of the prescribed hydrational potentialities was effected through the mutual sharing of the complements of H_2O^- units—a development which documented hydration as a bonding agent. A subsequent chapter of this book will be devoted to a consideration of hydrational bondage.

In contrast to the data for specific gravity at high solute concentrations the data for osmotic behavior at high solute concentrations, as was made evident in Table 6, evidenced a complete satisfaction of hydrational potentialities on an individual solute unit basis. This development naturally was subject to explanation as attributable to the fact that as ordinarily carried out in the laboratory there always was present in the bathing liquid an ample source of H_2O^- units to effect the complete hydration of all solute ions present. On this account in the procurement and analysis of osmotic data the volumetric increases on a standardized basis occasioned no surprise when they proved to be entirely independent of the concentration of the solute. This was true even when the concentration of the original solution inside the osmometer was so great as to preclude the complete satisfaction of the hydrational potentialities of the solute ions on an individual basis at the beginning of osmosis. There was this difference therefore : in concentrated solutions the specific gravity might attest the sharing of H_2O^- units by the involved solutes, whereas following osmosis in the presence of ample supplies of external water the hydrating ions possessed full complements of H_2O^- units. These developments represented osmosis as affording as definite a documentation of the integrity of the description of periodic hydrational potentiality as had been given by the specific gravity data of Chapter 3, and in many instances in which concentrated solutions were involved the osmotic data proved superior.

Naturally there was satisfaction in the indicated correlations, but the really surprising development in conjunction with researches on osmosis was the gradual accumulation of evidence that osmosis possessed potentialities for the dissociation of certain solute radicals whose integrity in aqueous solution repeatedly had been attested by specific gravity data. At the same time it became obvious that certain other solute radicals remained intact during osmosis. The situation in reality greatly extended the significance and usefulness of osmosis as a research tool. The development respecting the osmotic dissociation of solute radicals was considered to be of such merit as to be accorded special treatment in a subsequent chapter.

It seemed to be axiomatic that in aqueous solutions which involved element ions exclusively the osmotic behavior patterns of the ions would be analogous to those indicated for the Na^+ and Cl^- ions in Tables 6 and 7 of the preceding chapter except that different and characteristic hydrational potentialities would be prescribed. From this viewpoint a series of investigations was carried out designed to yield data affording a comparison of calculated or predicted standardized osmotic volumetric increases with values obtained as averages of replicated osmometer tests under controlled laboratory conditions. The results obtained were assembled and comprise Table 9.

In Table 9 the study of the osmotic behavior of $TlCl$ involved the use of a calibrated capillary tube because of the low solubility of the salt. The assumed ionic weight values given in the column headed Wa were prescribed by the description of periodic hydration and were used in preparing the desired molecular concentrations. It was obvious that had the concentrations been based upon the use of conventional atomic weight values greater osmotic increases in solution volume would have been obtained. The notable nicety of agreement between observed and calculated values was interpreted as attributable to several factors. One of these was the involvement of exceedingly large numbers of relatively independent solute units, representing conditions analogous to those involved in the elaboration of the gas laws. Another factor was the possession of a specific goal: on several occasions aberrant volumetric increases were found attributable to the use of salts which had been exposed to the air and had taken on moisture. Except for the possession of specific goals these errors would not have been detected. The basic and by far the most important factor naturally was the integrity of the description of periodic hydration,—an integrity documented by the observed specific gravity data given in Chapter 3.

Table 9

Data affording comparison of calculated and observed osmotic behavior patterns in validation of the description of periodic hydrational potentiality.

Solute	Assumed Ions	Wa	n	H	Wh	Vh	Va	Di	$\frac{Di}{2}$	$\frac{Va}{2}$	D	Calculated Standard Increases	Observed Standard Increase
LiCl	Li⁺	8	1	19	350	342	4	338	169	2	167	216.5	216
	Cl⁻	32	1	7	158	131	16	115	57.5	8	49.5		
KCl	K⁺	40	1	3	94	66	20	46	23	10	13	62.5	62.6
	Cl⁻	32	1	7	158	131	16	115	57.5	8	49.5		
NaF	Na⁺	24	1	11	222	200	12	188	94	6	88	215.5	215.3
	F⁻	16	1	15	285	271	8	263	131.5	4	127.5		
KF	K⁺	40	1	3	94	66	20	46	23	10	13	140.5	140.4
	F⁻	16	1	15	285	271	8	263	131.5	4	127.5		
KBr	K⁺	40	1	3	94	66	20	46	23	10	13	93.5	93.3
	Br⁻	68	2	12	284	229	34	195	97.5	17	80.5		
KI	K⁺	40	1	3	94	66	20	46	23	10	13	124.5	124.0
	I⁻	104	3	17	410	327	52	275	137.5	26	111.5		
NaBr	Na⁺	24	1	11	222	200	12	188	94	6	88	168.5	168.2
	Br⁻	68	2	12	284	229	34	195	97.5	17	80.5		
NaI	Na⁺	24	1	11	222	200	12	188	94	6	88	199.5	199.0
	I⁻	104	3	17	410	327	52	275	137.5	25	111.5		
RbF	Rb⁺	76	2	8	220	163	38	125	62.5	19	43.5	171.0	171.0
	F⁻	16	1	15	285	271	8	263	131.5	4	127.5		
RbCl	Rb⁺	76	2	8	220	163	38	125	62.5	19	43.5	93.0	93.0
	Cl⁻	32	1	7	158	131	16	115	57.5	8	49.5		
CsCl	Cs⁺	112	3	13	346	261	56	205	102.5	28	74.5	124.0	124.0
	Cl⁻	32	1	7	158	131	16	115	57.5	8	49.5		
TlCl	Tl⁺	164	4	10	344	233	82	151	75.5	41	34.5	84.0	84.0
	Cl⁻	32	1	7	158	131	16	115	57.5	8	49.5		

The solutes listed in Table 9 were of a type assumed to give rise to univalent element ions in aqueous solution. These were the solutes in which Kohlrausch had taken an intense interest in his later years, apparently in the hope of obtaining from electrometric data pertaining to the relative mobilities of the involved element ions a clue to the prevailing interrelationships. Without an adequate description of hydration his project was futile, but with the documentation of the integrity of the description of periodic hydrational potentiality it became obvious that this description could be further validated by electrometric data on solutions so prepared as to include appropriate allowances for the hydration of the ions and the consequent transfer of solvent to solute. Yet by this time it had become obvious that the osmotic behavior patterns afforded a much simpler and more inviting area for exploitation since, as noted, the involved procedures afforded a complete hydration of all solute ions at all concentrations of many solutes.

As documented by the data of Table 9, at atmospheric pressure and 25°C osmosis was a process mediated by the complete hydration of anhydrous solute ions within an absorptive membrane so arranged as to supply the involved H_2O^- units. A randomized exit of completely hydrated ions from the membrane took place at all liquid interfaces. Increases in internal solution volume effected by osmosis were attributable to the return of one-half the original number of solute ions to the liquid within the osmometer in the hydrated state, their original entry into the absorption membrane having involved the anhydrous state.

The derivation of the data given in Table 9 represented an exciting era in a long-continued search for observational data convincingly validating the basic order prescribed by the description of periodic hydrational potentiality and its corollaries. In the case of the classical data which many years ago had supplied the basis for the recognition of the behavior patterns documented in the gas laws, the remarkable nicety of correlation between observation and calculation or prediction had been attributed to the collective activity of statistically great numbers of individual entities each having a certain order of freedom of movement. In the case of the data given in Table 9, the hydrated solute entities also were present in statistically great numbers and possessed a certain order of freedom of movement. It was natural, therefore, to attribute the seemingly fantastic nicety of agreement between calculated or predicted behavior patterns and observed behavior patterns to the collective activity of hydrated solute ions. In numerous instances the actual concentrations of individual solute entities were in excess of those present in gases at the pressures which had been involved in the classical studies of the behavior patterns of gases.

One of the outstanding features of osmotic activity as represented by the data of Table 9 was the clearly evidenced complete hydration of all of the involved solute ions. Superficially one might examine the data of Tables 1 through 4 and venture to conclude that the mere act of dissolving the solutes in water had brought about a similar complete hydration of all of the involved solute units. Such a conclusion, however, was not justified. For one thing, it was found that although calculated and observed specific gravity values were in satisfactory agreement, the solute volumes calculated for the hydrated solute often were in excess of the total solution volumes. This development precluded the presence of completely hydrated individual solute ions, but documented the mutual sharing of the H_2O^- units and thus disclosed for the first time a mathematical basis for the appraisal of hydrational bondage. The subject was considered to be of sufficient interest and importance to merit special attention in a subsequent chapter of this book. For another thing, the calculated and observed specific gravity values for numerous substances in aqueous solution were found to be *not* in agreement when a complete hydration of all involved solute ions was assumed and postulated, yet were in agreement when alternative postulates were made. This was the situation noted for some of the solutes listed in Table 9.

From these considerations there evolved further studies whose results suggested that hydrational potentiality was concerned in the development of the solution attribute designated as acid-alkali reactivity. This subject will be taken up in a later chapter.

From the order of agreement between calculated and observed values in Table 9 it was concluded that osmosis had emerged as a reliable tool for research after many years of desultory involvement. That the description of periodic hydrational potentiality had mediated this emergence was very evident, as was also the fact that the opportunities for extending knowledge through the use of this tool were exciting, inviting and unlimited. Here was a broad new pathway to an improved understanding of solution phenomena, and since solutions included aspects pertaining to the gaseous and solid states the prospects were even more extensive. Although several chapters of this book will be concerned with the results of further researches dealing with osmosis it is to be recognized that these represent only feeble beginnings along a road leading to laws for solutions commensurable with the gas laws long esteemed as among the choicest descriptions in science.

CHAPTER 6

THE OSMOTIC BEHAVIOR OF NITRATE AND AMMONIUM RADICALS

Early attention to the osmotic behavior of the nitrate radical was prompted by a recognition of the fact that nitrates were more soluble in water than chlorides or their chemical analogs. It was obvious, therefore, that if an adequate interpretation of the osmotic behavior of the nitrate radical could be obtained, the list of cationic elements falling within the prescribed weight-range of the description of periodic hydrational potentiality could be appreciably extended. Such an extension to include heavy element ions seemed particularly desirable because of the obvious discrepancies between the conventional atomic weight values and the prescribed ionic weight values. The project was undertaken previous to the analyses of the specific gravity data for aqueous solutions of nitrates given in Table 4.

In the studies of the osmotic behavior patterns of nitrates it was found that the nitrate radical, appraised as NO_3^-, contributed to osmotically-mediated increases in solution volume precisely to the extent prescribed by the description of periodic hydrational potentiality for an element ion of the same weight. This development was responsible for the inclusion of a characterization for molecular ions in the description as given in Chapter 2. Data documenting the development have been assembled to comprise Table 10.

Such data as those given in Table 10 were interpreted as having been significant both in relation to the extension of the list of solute element ions whose osmotic behavior patterns were in validation of the description of periodic hydrational potentiality and in relation to the characterization of hydrational potentiality as conditioned by the nature of the involved solute unit. The development made it clear that the periodicity of hydrational potentiality and the periodicity of chemical attributes after Mendeleev were independent systems, even though in general they involved units with weights of the same order of magnitude. Moreover, there was the seeming implication that hydrational potentiality was of at least paramount importance since it had a primary relation to atomic nuclei and was more intimately integrated with the ninety-two naturally-occurring elements.

The data of Table 10 appeared to be of special interest in relation to involved heavy element ions. The fact that a consis-

Table 10

Data affording a comparison of calculated and observed osmotic behavior patterns in validation of the description of periodic hydrational potentiality.

Solute	Assumed Ions	Wa	n	H	Wh	Vh	Va	Di	$\frac{Di}{2}$	$\frac{Va}{2}$	D_2	u	Vi	Vt	O
NaNO₃	Na⁺	24	1	11	222	200	12	188	94	6	88	1	88		
	NO₃⁻	60	2	16	348	297	30	267	133.5	15	118.5	1	118.5	205.5	206
Mg(NO₃)₂	Mg⁺⁺	28	2	9	190	166	14	152	76	7	69	1	69		
	NO₃⁻	60	2	16	348	297	30	267	133.5	15	118.5	2	237	306	306
Al (NO₃)₃	Al⁺³	32	1	7	158	131	16	115	57.5	8	49.5	1	49.5		
	NO₃⁻	60	2	16	348	297	30	267	133.5	15	118.5	3	355.5	405	405
KNO₃	K⁺	40	1	3	94	66	20	46	23	10	13	1	13		
	NO₃⁻	60	2	16	348	297	30	267	133.5	15	118.5	1	118.5	131.5	131
Ca(NO₃)₂	Ca⁺⁺	44	1	1	62	36	22	14	7	11	-4	1	-4		
	NO₃⁻	60	2	16	348	297	30	257	133.5	15	118.5	2	237	233	233
Fe(NO₃)₃	Fe⁺³	58	2	17	364	314	29	285	142.5	14.5	128	1	128		
	NO₃⁻	60	2	16	348	297	30	267	133.5	15	118.5	3	355.5	483.5	483
Sr(NO₃)₂	Sr⁺⁺	80	2	6	188	132	49	92	45	20	26	1	26		
	NO₃⁻	60	2	16	348	297	30	257	133.5	15	118.5	2	237	253	263
AgNO₃	Ag⁺	95	3	21	474	394	43	345	173	21	149	1	149		
	NO₃⁻	60	2	16	348	297	30	267	133.5	15	118.5	1	118.5	257.5	267
BA(NO₃)₂	Ba⁺⁺	116	3	11	314	229	53	171	85.5	29	55.5	1	56.5		
	NO₃⁻	60	2	16	348	297	30	257	133.5	15	118.5	2	237	293.5	293
Hg(NO₃)₂	Hg⁺⁺	164	4	10	314	233	82	151	75.5	41	34.5	1	34.5		
	NO₃⁻	60	2	16	348	297	30	257	133.5	15	118.5	2	237	271.5	271
Pb(NO₃)₂	Pb⁺⁺	168	4	8	312	203	84	119	59.5	42	17.5	1	17.5		
	NO₃⁻	60	2	16	348	297	30	267	133.5	15	118.5	2	237	254.5	254

Supplementary Legend   Vt=Total Volume, or Standardized Volumetric Increase;
O=Observed Standardised Volumetric Increase.

tently uniform osmotic behavior pattern was evidenced for the nitrate ion supplied by correlation a documentation for the integrity of such ionic weight values as $Hg^{++}=164$ and $Pb^{++}=168$,—values also validated by the data for specific gravity given in Tables 1 and 4. It was recognized that in conjunction with hydration specific gravity and osmosis represented new approaches involving the behavior of individual units in a relatively moderate state of excitation. Nevertheless the ionic weight values cited for the Hg^{++} and Pb^{++} ions were so different from the conventional atomic weight values for these elements that they might seem to constitute an affront to chemical science. Although a subsequent chapter will be given over to a consideration of atomic weight, digression will be made at this point for an examination of the gravimetric data for oxides of mercury and lead,—data historically basic in the derivation of the conventional atomic weight values. These data as reported by FRANK W. CLARKE in an authoritative compendium on the atomic weights of the elements have been considered in Table 11.

In the data of Table 11 the gravimetric ratios in the right hand column when interpreted to the base $O=16$ yielded conventional atomic weight values of the order of $Hg=200.61$ and $Pb=207.21$. It was to be noted, however, that these observed ratios when interpreted to ionic bases prescribed by the description of periodic hydrational potentiality yielded ionic weight values commensurate with the ionic weight values documented as valid by the appropriate data of specific gravity and osmosis. In Table 2 the specific gravity of tungsten trioxide in aqueous solution evidenced a weight of 12 for the O^{--} ion in the anhydrous state. In the chapter which follows, data for the osmotic behavior of the O^{--} ion evidence for it a weight of 12 in the anhydrous state.

It was recognized that the observed osmotic behavior of the nitrate radical, however helpful in conjunction with the further validation of the description of periodic hydrational potentiality, actually had introduced another problem. In crude extracts of organic tissues nitrate radicals often have been detected although their internal presence has not been evidenced as a prerequisite to growth and development. However, no instance was known in organic chemistry in which the NO^-_3 radical existed as an intact unit in a proteinaceous molecule. It followed, therefore, that incident to protoplasmic metabolism the NO^-_3 radical must undergo dissociation. In general such a projected dissociation within a protoplasmic matrix might be attributed to the activity of catalysts of an organic nature, designated as enzymes. Subsequent researches on osmosis, however,

Table 11

Data affording a comparison of calculated and observed gravimetric ratios in validation of the description of periodic hydrational potentiality.

Chemical Compound	Assumed Ions	Per cent heavy element present calculated by use of the description of periodic hydrational potentiality			Per cent heavy element present as calculated from gravimetric data of chemical science
		By indicated ionic weights	By indicated combining weights	Average values	
HgO	Hg^{++} O^{--}	93.18 $Hg^{++}=164$ $O^{--}=12$	92.04 $Hg^{++}=162$ $O^{--}=14$	92.61	92.595 ± .0003*
PbO	Pb^{++} O^{--}	93.33 $Pb^{++}=168$ $O^{--}=12$	92.22 $Pb^{++}=166$ $O^{--}=14$	92.97	92.827 ±.0013**

*Average of 15 determination by TURNER, ERDMANN and MARCHAND and HARDEN.
**Average of 9 determinations by BERZELIUS.

led to wholly unexpected results and an alternative viewpoint. The developments were considered to be of such importance as to merit the special treatment accorded them in a later chapter.

The success which had attended the study of the osmotic behavior of the nitrate radical led naturally to a similar study of the ammonium radical. At the outset there arose uncertainty as to the validity of the conventional appraisal of the constitution of the radical. The description of periodic hydrational potentiality prescribed a weight of 2 for the atom of hydrogen in the neutral state, whereas conventional chemical science interpreted hydrogen as having an atomic weight of 1.008 without regard to its state. Disregarding the fraction the conventional weight of the NH^+_4 radical would be 18; the prescribed weight of the NH^+_4 radical would be 24 : the prescribed weight of a projected radical NH^+_2 would be 20. Under these circumstances it seemed appropriate to base the study of the osmotic behavior of ammonium chloride on each of two hypotheses (1) that the ammonium radical was of the composition NH^+_2 and weighed 20, and (2) that the ammonium radical was of the composition NH^+_4 and weighed 24. Solutions of ammonium chloride were prepared with concentrations based on each of these values. The results obtained were assembled to comprise Table 12.

An examination of the results given in Table 12 made the following items seem altogether likely : (1) that the composition of the ammonium radical was NH_4 as conventionally appraised; (2) that the weight of each of the four hydrogen atoms in the neutral state was 2 as prescribed by the description of periodic hydrational potentiality; and (3) that the NH^+_4 radical had remained intact during osmosis and had become hydrated to the precise extent prescribed by the description of hydration for an element ion of the same weight.

With the assurance of integrity for the ammonium radical afforded by the foregoing results the studies of osmotic behavior were extended. The results obtained were assembled to comprise Table 13 and for continuity the data which had been obtained for ammonium chloride were included.

The data given in Table 13 were interpreted as substantiating the conclusions ventured by the examination of the data of Table 12. Up to a point, therefore, the results obtained were satisfying, yet as was the case with the nitrate radical, there was no known instance in which the ammonium radical remained intact and entered as such into the composition of an organic molecule. In some manner, therefore, protoplasmic metabolism included the ultimate partial or complete dissociation of the NH^+_4 radical. The problem became a

Table 12

Data affording a comparison of calculated and observed osmotic behavior patterns of ammonium chloride in relation to the composition and weight of the ammonium radical. Calculated values, column Vt; observed values, Column O.

Solute interpreted as	Assumed Ions	Wa	n	H	Wh	Vh	Va	Di	$\frac{Di}{2}$	$\frac{Va}{2}$	D_2	u	Vi	Vt	O
NH$_4$Cl 52 gms. per liter	NH^+_2	20	1	13	254	236	10	226	113	5	108	1	108		
	Cl	32	1	7	158	131	16	115	57.5	8	49.5	1	49.5	157.5	127.6
NH$_4$Cl 56 gms. per liter	NH^+_4	24	1	11	222	200	12	188	94	6	88	1	88		
	Cl	32	1	7	158	131	16	115	57.5	8	49.5	1	49.5	137.5	137.5

Table 13

Data affording a comparison of calculated and observed osmotic behavior patterns for some ammonium compounds. Calculated values, Column Vt ; observed values, Column O.

Solute	Assumed Ions	Wa	n	H	Wh	Vh	Va	Di	$\frac{Di}{2}$	$\frac{Va}{2}$	D_2	u	Vi	Vt	O
NH$_4$F	NH^+_4	24	1	11	222	200	12	188	94	6	88	1	88		
	F^-	16	1	15	285	271	8	263	131.5	4	127.5	1	127.5	215.5	215.5
NH$_4$Cl	NH^+_4	24	1	11	222	203	12	188	64	6	88	1	88		
	Cl^-	32	1	7	158	131	16	115	57.5	8	49.5	1	49.5	137.5	137.5
NH$_4$NO$_3$	NH^+_4	24	1	11	222	230	12	188	94	6	88	1	88		
	NO^-_3	60	2	16	348	297	30	267	133.5	15	118.5	1	118.5	205.5	206.5
NH$_4$Br	NH^+_4	24	1	11	222	200	12	188	94	6	88	1	88		
	Br^-	68	2	12	284	229	34	195	97.5	17	80.5	1	80.5	168.5	168.5
NH$_4$I	NH^+_4	24	1	11	222	200	12	188	94	6	88	1	88		
	I^-	104	3	17	410	327	52	275	137.5	26	111.5	1	111.5	199.5	199.5

particularly challenging one because the hydrogen components of the NH^+_4 radical were subject to projection as having entered the chemical association as anionic H^- ions. The description of periodic hydrational potentiality prescribed for such H^- ions a weight of zero, a combining weight of 1.0, and an ability to take on 23 H_2O^- units in hydration. It followed that the release of such H^- ions from the NH^+_4 radical and their subsequent behavior would have an important bearing on the conventional appraisal of the atomic weight of hydrogen as well as upon the further validation of the integrity of the description of periodic hydrational potentiality. The results of researches interpreted as involving the complete dissociation of nitrate and ammonium radicals have been considered in a subsequent chapter.

Collectively the data of Tables 10, 12 and 13 were interpreted as having demonstrated that the nitrate radical NO^-_3 and the ammonium radical NH^+_4 remained intact in osmosis restricted to individual chemical compounds and that as intact radicals they possessed the potentialities for hydration prescribed by the description of periodic hydrational potentiality for element ions of the same weight in the anhydrous state. On the basis of this evidence an appropriate change was made in the description, originally derived from and restricted to the behavior of element ions.

The foregoing interpretation was considered to be a very important development because it was recognized that if the behavior pattern was representative for intact solute radicals there was at hand a hitherto unavailable significance for osmosis as an index of the nature and status of solute units, even when the dissolved substances included solute units of considerable complexity. At the very least osmotic behavior thus supplied a means of testing the validity of assumed ionic states present in the solution before osmosis. Moreover, as evidenced by the context of the chapter which follows, osmosis afforded a means of detecting and analyzing any changes of solute nature and status which might take place incident to osmosis.

LITERATURE CITED

CLARKE, FRANK WIGGLESWORTH (1922). A Recalculation of the Atomic Weights. (Fourth edition, revised and enlarged). Memoirs of the National Academy of Sciences 16:5-418.

CHAPTER 7

THE DISCOVERY OF OSMOTIC DISSOCIATION
POTENTIALITIES

As noted previously, the study of the osmotic behavior of nitrates
was prompted only by a recognition of their general solubility and
consequent potential usefulness in extending the study of ions. Yet
when the nitrate and ammonium radicals became evidenced as having remained intact in osmosis and as having hydrational potentialities precisely those prescribed for element ions of the same weight
there was the prospect of utilizing solute radicals in the further
validation of the description of periodic hydrational potentiality in a
manner analogous to that involved in the data of Tables 3 and 4 relating to specific gravity. In keeping with this viewpoint the
osmotic behavior patterns of some sulfates, phosphates and carbonates were studied. The results obtained have been given in Table
14.

As indicated in the data of Table 14 the assumptions made with
respect to the sulfates, phosphates and carbonates involved the concept of solute radicals remaining intact throughout the osmotic process, as had been evidenced in the behavior of the nitrate and ammonium radicals. Yet it was obvious from an examination of the data
that the observed values, Column O, were not in agreement with the
values derived through the use of the indicated assumptions as given
in Column Vt. Under the circumstances there seemed to be no
alternative but to venture that something had happened to the radicals and to make empirical calculations with a view to the possible
derivation of a satisfactory explanation of the developments. Such
calculations eventually led to the interpretation indicated in the
comparisons made available from the data of Table 15.

An examination of the data of Tables 14 and 15 seemed to lead
inevitably to the conclusion that the observed osmotic increases in
solution volume had been brought about by the atomic components
of the originally assumed radicals (data of Table 14) acting as element ions (data of Table 15). This conclusion was considered to be
not only in contrast to the appraisal of the data for the osmotic behavior of the nitrate and ammonium radicals, but also in contrast to
commonly held appraisals of stability based upon patterns of chemical behavior. There remained the important question of the stability of the element ions released into osmotized solutions and this

Table 14

Data affording a comparison of calculated and observed osmotic behavior values for some solutions assumed to contain intact radicals. Calculated values, Column Vt: Observed values, Column O.

Solute	Assumed Ions	Wa	n	H	Wh	Vh	Va	Di	$\frac{Di}{2}$	$\frac{Va}{2}$	D_2	u	Vi	Vt	O
Li₂SO₄	Li⁺	8	1	19	350	342	4	338	169	2	167	2	334	503	918
	SO⁻⁻	92	3	23	506	430	46	384	192	23	169	1	169		
Na₂SO₄	Na⁺	24	1	11	222	200	12	188	94	6	88	2	176	345	760
	SO₄⁻⁻	92	3	23	506	430	46	384	192	23	169	1	169		
K₂SO₄	K⁺	40	1	3	94	66	20	46	23	10	13	2	26	195	610
	SO₄⁻⁻	92	3	23	506	430	46	384	192	23	169	1	169		
Na₂HPO₄	Na⁺	24	1	11	222	200	12	188	94	6	88	2	176		
	H⁺	4	1	21	382	378	2	376	188	1	187	1	187	362	957
	PO₄⁻³	88	2	2	124	72	44	28	14	22	−8	1	−8		
K₂HPO₄	K⁺	40	1	3	94	66	20	46	23	10	13	2	26		
	H⁺	4	1	21	382	378	2	376	188	1	187	1	187	205	807
	PO₄⁻³	88	2	2	124	72	44	28	14	22	−8	1	−8		
Na₂CO₃	Na⁺	24	1	11	222	200	12	188	94	6	88	2	176	313.5	725
	CO₃⁻⁻	56	2	18	380	331	28	303	151.5	14	137.5	1	137.5		
K₂CO₃	K⁺	40	1	3	94	66	20	46	23	10	13	2	26	163.5	575
	CO₃⁻⁻	56	2	18	380	331	28	303	151.5	14	137.5	1	137.5		

Table 15

Data affording a comparison of the observed osmotic behavior of some sulfate, phosphate and carbonate solutions with the osmotic behavior calculated on the assumption that the involved radicals underwent complete osmotic dissociation.

Calculated values, Column Vt. Observed values, Column O.

Solute	Assumed Ions	Wa	n	H	Wh	Vh	Va	Di	Di/2	Va/2	D₂	u	Vi	Vt	O
Li$_2$SO$_4$	Li$^+$	8	1	19	350	342	4	338	169	2	167	2	334		
	S^{+8}	44	1	1	62	36	22	14	7	11	−4	1	−4		
	O^{--}	12	1	17	318	305	6	300	150	3	147	4	588	918	918
Na$_2$SO$_4$	Na$^+$	24	1	11	222	200	12	188	94	6	88	2	176		
	S^{+8}	44	1	1	62	36	22	14	7	11	−4	1	−4		
	O^{--}	12	1	17	318	306	6	300	150	3	147	4	588	760	760
K$_2$SO$_4$	K$^+$	40	1	3	94	66	20	46	23	10	13	2	26		
	S^{+8}	44	1	1	62	36	22	14	7	11	−4	1	−4		
	O^{--}	12	1	17	318	306	6	300	150	3	147	4	588	610	610
Na$_2$HPO$_4$	Na$^+$	24	1	11	222	200	12	188	94	6	88	2	176		
	H$^+$	4	1	21	382	365	2	363	181.5	1	180.5	1	180.5		
	P^{+5}	40	1	3	94	66	20	46	23	10	13	1	13		
	O^{--}	12	1	17	318	306	6	300	150	3	147	4	588	957.5	957
K$_2$HPO$_4$	K$^+$	40	1	3	94	66	20	46	23	10	13	2	26		
	H$^+$	4	1	21	382	365	2	363	181.5	1	180.5	1	180.5		
	P^{+5}	40	1	3	94	66	20	46	23	10	13	1	13		
	O^{--}	12	1	17	318	306	6	300	150	3	147	4	588	807.5	807
Na$_2$CO$_5$	Na$^+$	24	1	11	222	200	12	188	94	6	88	2	176		
	C^{+4}	20	1	13	254	236	10	226	113	5	108	1	108		
	O^{--}	12	1	17	318	306	6	300	150	3	147	3	441	725	725
K$_2$CO$_3$	K$^+$	40	1	3	94	66	20	46	23	10	13	2	26		
	C^{+4}	20	1	13	254	236	10	226	113	5	108	1	108		
	O^{--}	12	1	17	318	306	6	300	150	3	147	3	441	575	575

question would have to be answered through further research. Yet the existence of the element ions within the absorptive membrane matrix seemed assured, entailing presumptively a consequent availability for roles in protoplasmic metabolism.

Within the area of chemistry it was obvious that the projected and evidenced characterizations of such element ions and ionic weight values as $H^+=4$, $O^{--}=12$, $C^{+4}=20$, $P^{+5}=40$ and $S^{+6}=44$ brought forward a new and unexpected validation of the description of periodic hydrational potentiality and its corollaries. This development alone had dramatic significance, since it had an intimate bearing on the evolution of chemical science in relation to the use of the value $O=16$. Constructively for chemical science the development projected a new appreciation of order at the atomic and molecular level. It also projected a new technique for the dissociation of chemical compounds, replete with a provision for the diagnosis of atomic composition. There was included a prospect for usefulness in the isolation of desired sub-units.

Although the discovery of osmotic dissociation potentialities took place in the area of inorganic solutes there appeared to be no reason why its sphere of usefulness should not extend into the area of organic solutes. Such an extension had special appeal. Except for the element nitrogen all of the elements essential to organic growth and development potentially were made available without projected recourse to enzymatic activity whenever osmotically-active vacuoles were present in a suitable matrix and substrate. Clearly the osmotic behavior of organic solutes had been brought forward and emphasized as a research area of great interest, promise and challenge.

For more immediate attention, however, there was supplementary research in the area of inorganic solutes. There were numerous other compounds assumed to yield oxygen-containing radicals in aqueous solution and hence more or less analogous to the sulfates, phosphates and carbonates which had been evidenced as subject to osmotic dissociation. A study of such compounds led to the results given in Table 16.

The results given in Table 16 not only supplemented the data of Table 15 in documenting the existence of osmotic potentialities for dissociation but also supplied evidence relating specifically to the weights of some heavy element ions in both the anhydrous and hydrated states. In the latter capacity they extended the observational data substantiating the validity of the description of periodic hydration and its corollaries with the anhydrous ionic weight values $Se^{+4}=76$, $Se^{+6}=80$, $Mo^{+6}=96$, $Te^{+6}=116$ and $W^{+6}=160$. In an

Table 16

Data interpreted as evidence of the complete osmotic dissociation of some radicals containing heavy element ions. Calculated values, Column Vt. Observed values, Column O.

Solute	Assumed Ions	Wa	n	N	Wh	Vh	Va	Di	$\frac{Di}{2}$	$\frac{Va}{2}$	D_2	u	Vi	Vt	O
Na$_2$SeO$_3$	Na$^+$	24	1	11	222	200	12	188	94	6	88	2	176		
	Se^{+4}	76	2	8	220	163	38	125	62.5	19	43.5	1	43.5		
	O^{--}	12	1	17	318	306	6	300	150	3	147	3	441	660.5	660
Na$_2$SeO$_4$	Na$^+$	24	1	11	222	200	12	188	94	6	88	2	176		
	Se^{+6}	80	2	6	188	132	40	92	46	20	26	1	26		
	O^{--}	12	1	17	318	305	6	300	150	3	147	4	588	790	790
(NH$_4$)$_2$MoO$_4$	NH$_4^+$	24	1	11	222	200	12	188	94	6	88	2	176		
	Mo^{+6}	96	3	21	474	394	48	346	173	24	149	1	149		
	O^{--}	12	1	17	318	306	6	300	150	3	147	4	588	913	913
Na$_2$TeO$_4$	Na$^+$	24	1	11	222	200	12	188	94	6	88	2	176		
	Te^{+6}	116	3	11	314	229	58	171	85.5	29	56.5	1	56.5		
	O^{--}	12	1	17	318	306	6	300	150	3	147	4	588	820.5	820
Na$_2$WO$_4$	Na$^+$	24	1	11	222	200	12	183	94	6	88	2	176		
	W^{+6}	160	4	12	376	264	80	184	92	40	52	1	52		
	O^{--}	12	1	17	318	306	6	200	150	3	147	4	588	716	716

analogous manner the data of Table 15 had supplied additional ionic weight values for a series of light elements : $H^+=4$, $O^{--}=12$, $C^{+4}=20$, $P^{+5}=40$ and $S^{+6}=44$. It was thus made evident that by virtue of the osmotically mediated dissociation of solute radicals element ions hitherto in practice inaccessible had been made subject to characterization.

The developments incited speculation as a prelude to further research. It was assumed that the involved membrane represented a uniform degree of osmotic stress and that the bondages involved in solute radicals represented different degrees of stress. In conformity with the latter assumption it appeared reasonable to venture that some of the element ions evidenced as released through osmotic dissociation would possess potentialities for dissociation which would exceed the more common in vitro potentialities of solutes. Thus there was envisioned the possibility that in mixtures of solutions subjected to osmotization one or more of the ions released through osmotic dissociation might bring about the dissociation of solute radicals otherwise remaining intact. The possibility was of special interest in relation to the nitrate and ammonium radicals considered in Chapter 6.

Another prospective advantage accruing from the discovery of osmotic potentialities for dissociation was its relation to the area of diagnosis. In the data of Tables 15 and 16 the nicety of agreements between calculated and observed values attested a validity for the assumed complete dissociation and atomic constitution. It was obvious, however, that had osmosis effected only a partial dissociation of the radicals or had the assumed atomic composition of the radicals been erroneous, no agreement between calculated and observed values would have been obtained. The situation thus emphasized the potential value of osmotic behavior as an index of radical composition and degree of dissociation: both had to be correct to permit any meticulous correlation.

Although the documenting data of Tables 15 and 16 involved inorganic solutes the potential index value of osmotic behavior held much of promise in relation to an improved understanding of the behavior patterns of organic solutes. It appeared reasonable to assume that organic solutes would be subject to osmotically-mediated hydration and perhaps also to osmotically-mediated dissociation, partial or complete. With respect to the osmotic behavior of solute non-electrolytes the situation remained particularly formidable and challenging. For inorganic solutes the osmotic data consistently had evidenced for solutions at all concentrations an absence of nondissociated neutral solute molecules, and there was an allied implication that hydration was exclusively an attribute of ions.

The discovery of the existence of osmotic potentialities for the dissociation of solute radicals brought forward for consideration some data relating to ions which under usual circumstances had not been sufficiently stable to permit meticulous appraisal, on which account it seemed appropriate to make a brief summary of osmotic behavior as an index of hydration in a manner analogous to the summary given in Table 5 for specific gravity as an index of hydration. The osmotic data tabulated to this point have been represented thus in the following Table 17.

The compilation of data given in Table 17 was subject to appraisal as a supplement to the compilation given in Table 5. The data of Table 5, representing correlations between predicted or calculated specific gravity values and values obtained by various investigators, were interpreted as having documented a validity for the involved description of periodic hydrational potentiality. Because of osmotic dissociation the data of Table 17 emphasized in particular the behavior patterns of ions within the first period.

An important diagnostic feature of osmotic dissociation was the period of time involved in the attainment of maximum increases in solution volume. In the absence of osmotic dissociation maximal increases commonly were attained in somewhat less than 48 hours when the solute concentrations were of the order of one molecular. In contrast, the maximal increases in solution volume in the case of sucrose were not attained until a period of about 168 hours had elapsed. The data of Table 8 include relative mobility values for anhydrous and hydrated ions which have been sustained repeatedly by the relative time requirements involved in the completion of osmotic activity.

The discovery of the existence of osmotic potentialities for the partial or complete dissociation of molecular solutes was appraised as a development opening up new avenues of approach in research. There was envisioned the probability that hitherto unknown or difficultly available substances might be derived through partial or selective osmotic dissociation. It seemed likely that osmosis would become of service in analyses of the effects of a wide variety of toxins, depressants and stimulants, in part through their effect on the rate of osmotic activity interpreted as attributable to modifications of the involved membrane, and in part through the osmotic effect produced by the agents alone. There has been some evidence, for example, that some effects of antibiotics have been attributable

Table 17

Data for the solute ions evidenced by osmotic behavior in preceding tables as having been hydrated to the extent indicated. Arrangement in periods.

First Hydration Period			Second Hydration Period			Third Hydration Period			Fourth Hydration Period		
Ions	Wa	H	Ions	Wa	H	Ions	Wa	H	Ions	Wa	H
H^+	4	21	Fe^{+3}	58	17	Ag^+,Mo^{+6}	96	21	W^{+6}	160	12
Li^+	8	19	NO^-_3	60	16	I^-	104	17	Te^+,Hg^{++}	164	10
O^{--}	12	17	Br^-	68	12	Cs^+	112	13	Pb^{++}	168	8
F^-	16	15	Rb^+	76	8	Ba^{++},Te^{+6}	116	11			
C^{+4}	20	13	Sr^{++}	80	6						
Na^+,NH^+_4	24	11									
Mg^{++}	28	9									
Al^{+3},Cl^-	32	7									
P^{+5},K^+	40	3									
S^{+6},Ca^{++}	44	1									

to an action on the membranes of the intestinal tract. There was the prospect for the production of improved osmotic membranes and increases in both quantitative and qualitative dissociation potentialities in the direction of selectivity.

In the inorganic area it was clear that some atomic ions evidenced as subject to release through osmotic dissociation would have potentialities for chemical reactivity presently obscure or recognized and attributed to enzymes. For the years ahead it seemed quite conservative to hold that osmotic dissociation potentialities would greatly enhance the selectivity of procurement for desired reactants and hence appreciably extend the degree of control over both dissociative and synthetic processes. For the present there was satisfaction that osmotic dissociation had supplied an unexpected confirmation of the specific change in weight with ionization which had been prescribed by the description of periodic hydrational potentiality and its allied supplementary validation of the description in its entirety.

CHAPTER 8

ON THE RELEASE OF ATOMIC NITROGEN IONS

The context of Chapter 6 included data which made it quite clear that in osmosis the solute radicals NO_3^- and NH_4^+ behaved in a manner precisely that which would have characterized element ions of corresponding weights. In one way this was considered as an important circumstance, since it supplied a basis for the calculation of the osmotic behavior of solutes giving rise to solute radicals which remained intact during osmosis. It also supplied a basis for the detection of osmotic dissociation potentialities, as was made evident in the data given in Chapter 7. Yet in another way the osmotic behavior of the nitrate and ammonium radicals was considered as frustrating and challenging. It was recognized that whereas the nitrate and ammonium radicals commonly played important roles in plant nutrition there was no incorporation of these radicals in the elaboration of proteins. It seemed to follow that protoplasmic metabolism relating to nitrogen must commonly involve the dissociation of nitrate and ammonium radicals by a process not exclusively osmotic. On occasion the presence of solute NO_3^- and NH_4^+ radicals had been noted in cell sap, a circumstance which was corroborative of the intact status noted as persistent throughout osmosis. The internal dissociation of the nitrate and ammonium radicals conditionally was subject to appraisal as brought about by enzymes. Yet such an appraisal became subject to critical review when osmosis became recognized as possessing potentialities for the dissociation of numerous solute radicals as indicated in the data of Chapter 7. From the nutritional viewpoint it became obvious that such fundamentally important elements as oxygen, hydrogen, sulfur, carbon and phosphorus were capable of being released osmotically from appropriate solute radicals. It seemed untenable to hold that element ions of nitrogen were not made available for protein synthesis.

The foregoing considerations appeared to suggest that since osmosis had demonstrably evidenced dissociation potentialities in relation to sulfate, phosphate and carbonate radicals, it was possible that the element ions released through such dissociations by osmosis might catalyze the dissociation of nitrate and ammonium radicals. As thus projected it was obvious that the released element ions, if effective, might be subject to characterization as enzymes; but at least their identity might become established.

The indicated suggestion was followed by a rather extensive series of tests in which paired solutes were osmotized and the volumetric increases obtained were compared with the volumetric increases subject to calculation as summations of the values predictable for the solutes osmotized as individuals. These tests yielded results which were considered as significant only as evidence that no interaction took place between the involved solutes, the total volumetric increases obtained having been the sums of the values characteristic for the individual solutes. Indirectly the results afforded further proof of the integrity of the description of periodic hydrational potentiality, but their presentation in such a role did not appear justifiable following the extensive tabulations in preceding chapters.

After what seemed to be a dull routine of failures it was found that when a mixture of copper sulfate and potassium nitrate was osmotized the volumetric increase obtained was greater than that predictable as the sum of the increases of the components osmotized separately. When it was found that the volumetric increase was precisely that which was subject to calculation on the assumption that the nitrate radicals had become dissociated and that the element ions thus released had become completely hydrated in accordance with the prescription provided by the description of periodic hydration it was concluded that the nitrate radical had been completely dissociated and that thus there had been released atomic nitrogen ions.

Following upon a succession of failures the evidenced dissociation of the nitrate radical was a welcome development. Tests were carried out in which on a molecular basis the ratios of KNO_3 and $CuSO_4$ in the mixtures to be osmotized were varied in the order of simple whole numbers. These tests indicated that on a molecular basis $CuSO_4$ could bring about the dissociation of four nitrate radicals,—a relationship which was interpreted as suggesting a reversal of charge on the copper. Tests were carried out with mixtures of KNO_3 and $Cu(NO_3)_2$. The volumetric increases obtained indicated that the NO_3^- radicals under such conditions remained intact, and it was concluded that the Cu^{++} ion alone was not capable of bringing about the dissociation of nitrate radicals. The osmotic behavior of copper sulfate alone had evidenced a complete dissociation of the sulfate radical, followed by a complete hydration of the released element ions. The nature of the activity involved in the evidenced breakdown of the nitrate radical and the agent or agents supplementing the activity of the Cu^{++} ion were not determined and thus remained conjectural. That the activity was not subject to appraisal as chemical within the common category seemed evidenced by the results of subsequent tests with NH₄Cl and CuSO₄.

When mixtures of ammonium chloride and copper sulfate were osmotized the volumetric increases obtained exceeded the sums of the values characterizing the osmotic activity of the substances as individuals. Moreover, as with the nitrate-sulfate mixture, the volumetric increases obtained were precisely those subject to calculation on the assumption that a complete dissociation of the ammonium radical had taken place and that the atomic ions thus released had become completely hydrated in accordance with the arrangement prescribed by the description of periodic hydrational potentiality. Further tests with mixtures involving simple molecular ratios of the components gave results which indicated that a molecule of CuSO₄ could effect the decomposition of four ammonium radicals in a manner analogous to reactivity with the nitrate radical. Since the nitrate and ammonium radicals bore opposite charges the involved reactivity appeared to be not in conformity with the more common type of chemical reaction.

The evidenced dissociation of the ammonium radical was interpreted as a development of unusual interest and potential significance because it gave characterization to negatively-charged atomic ions of hydrogen. The nitrogen atoms which were evidenced as released were polyvalent cationic N^{+5} ions whose behavior patterns were identical with those of the N^{+5} ions evidenced as released through the dissociation of the nitrate radical, as was to be expected. The O^{--} atomic ions evidenced as released through the dissociation of NO^-_3 radicals exhibited the same behavior patterns as the O^{--} atomic ions evidenced as released through the osmotic dissociation of sulfates, phosphates and carbonates,—as was to be expected. On the other hand the atomic H^- ions evidenced as released through the dissociation of NH^+_4 radicals were in sharp contrast to the H^+ ions evidenced as released through the osmotic dissociation of acidic salts. As prescribed by the description of periodic hydrational potentiality the weight of an H^- ion was subject to calculation or appraisal as zero; but zero weight was not an attribute readily subject to measurement or even detection by the conventional procedures of chemical science. The prescribed combining weight of an H^- ion, however, was 1.0, and this attribute was amenable to measurement. It seemed altogether reasonable to conjecture that this attribute often had been measured in chemical science.

It was obvious that the hydrational potentiality prescribed for an

H⁻ ion of zero weight was maximal at 23 H_2O^- units and it followed that the release of four H⁻ atomic ions through the dissociation of the NH_4^+ radical afforded through the medium of volumetric increase incident to hydration an indirect method of attesting zero weight. The development was interpreted as being of important significance within the immediate area of solute ions. The osmotic behavior of the H^+ ion had been such as to evidence volumetric increases subject only to correlation with the prescribed weight of 4. Since the osmotic behavior of the H^- ion was such as to evidence volumetric increases subject only to correlation with the prescribed weight of zero it followed that the weight of a neutral atom of hydrogen was subject to correlation with the weight 2 as prescribed by Corollary 1 of the description of periodic hydration.

As appraised in the foregoing considerations the molecular weight of biatomic hydrogen gas became subject to evaluation as 4, a value in contrast to that of contemporary chemical science. This situation made imperative a special study of the behavior patterns of gaseous hydrogen since it was obvious that as solute ions the element conformed nicely to the behavior patterns prescribed by the description of periodic hydrational potentiality. The results obtained in this special study not only made it clear that the molecular weight of biatomic hydrogen gas was 4 and not 2 as conventionally held, but also afforded an explanation of how it came about that a weight of 2 was derived. The data will be considered in detail in subsequent chapters.

In conjunction with a projected application of the Graham Law of Diffusion the osmotic evidence that the H^- ion possessed a weight of zero was of interest far beyond the immediate area of solution phenomena. For more than half a century it had been recognized that dissolved substances, especially in dilute solution, behaved in such a manner as to evidence an appreciable degree of freedom of movement for the involved solute units. For solutions of electrolytes this degree of freedom had been memorialized in the Kohlrausch law of the independent migration of ions. In general it had been broadly memorialized in the acknowledged analogy between the behavior patterns of solute units and gases. The data given in Chapter 3 relating to specific gravity as an index of hydration had been restricted to solutions in which all solute ions had been indicated as hydrated, since the index was used as a means of validating periodic hydrational potentialities. The same index might be used to document situations in aqueous solutions of certain substances in which some specific ions were present in the hydrated state accompanied by specific ions in the anhydrous state. Special consideration will be given to such situations in a subsequent chapter. At this point, however, it will suffice to note the intensified analogy which obtains between solute units and gases when solute units are present in the anhydrous state. The description of periodic hydrational potentiality prescribed hydration as restricted to ions of the weight range O-184 on the $O_2 = 32$ scale, on which account it followed that an ion such as U^{+6} would have a weight exceeding the range and would remain anhydrous. The formation of the uranyl cationic radical might be cited as in conformity with the prescribed status. The behavior of the U^{+6} ion will be considered further at a later point.

The basic observational data interpreted as having documented the release of atomic nitrogen ions have been given in Tables 18 and 19.

The results given in Tables 18 and 19 represented the averages obtained in replicated tests in which the departures from the values calculated for complete dissociation of the involved radicals were negligible. It was concluded, therefore, that incident to osmotic activity copper sulfate was capable of bringing about the complete dissociation of nitrate and ammonium radicals to effect the release of atomic nitrogen ions.

It had become obvious in conjunction with previous research results that chemical compounds were units held together by bondages representing different degrees of tenacity. The bondages of substances insoluble in water represented linkages not subject to breakage by the dissociating potentialities of an aqueous solvent. The bondages of substances yielding solute radicals in water, or including same, represented linkages partially subject to breakage. With respect to the nitrate and ammonium radicals it was obvious that individually these radicals represented linkages which could not be broken by osmosis alone. That these linkages could be broken by the synergic or catalytic action of specific accompanying solute ions appeared to have deep-seated potential significance in the direction of an improved understanding of metabolic processes. Since all soluble sulfates examined had been evidenced as subject to a complete osmotic dissociation of the sulfate radical it seemed obvious that in any mixture containing Cu^{++} ions and SO_4^{--} radicals osmotization would mediate a potentiality for the release of atomic nitrogen ions from either nitrate or ammonium radicals.

The results appeared to be of special interest in relation to metabolism involving nitrogenous compounds and prompted a study of urea. The results obtained will be taken up in a later chapter.

Table 18

Data relating to the osmotic behavior of mixtures of KNO_3 and $CuSO_4$, interpreted as evidencing the complete dissociation of the nitrate radical.

Solutes	Assumed Ions and Wa	n	H	Wh	Vh	Va	Di	$\frac{Di}{2}$	$\frac{Va}{2}$	Vi	u	V_2	Vt	O
KNO_3	K^+ =40	1	3	94	66	20	46	23	10	13	1	13		
	NO^-_3 =60	2	16	348	297	30	267	133.5	15	118.5	1	118.5		
+														
$CuSO_4$	Cu^{++} =62	2	15	332	280	31	249	124.5	15.5	109	1	109		
	S^{+6} =44	1	1	62	36	22	14	7	11	-4	1	-4		
	O^{--} =12	1	17	318	306	6	300	150	3	147	4	588	824.5	1235
KNO_3	K^+ =40	1	3	94	66	20	46	23	10	13	1	13		
	N^{+5} =24	1	11	222	200	12	188	94	6	88	1	88		
	O^{--} =12	1	17	318	306	6	300	150	3	147	3	441		
+														
$CuSO_4$	Cu^{++} =62	2	15	332	280	31	249	124.5	15.5	109	1	109		
	S^{+1} =44	1	1	62	36	22	14	7	11	-4	1	-4		
	O^{--} =12	1	17	318	306	6	300	150	3	147	4	588	1235	1235

Table 19

Data relating to the osmotic behavior of mixtures of NH_4Cl and $CuSO_4$ interpreted as evidencing dissociation of the ammonium radical.

Solutes	Assumed Ions and Wa	n	H	Wh	Vh	Va	Di	$\dfrac{Di}{2}$	$\dfrac{Va}{2}$	Vi	u	V_2	Vt	O
NH_4Cl	NH_4^+ =24	1	11	222	200	12	188	94	6	88	1	88		
	Cl^- =32	1	7	158	131	16	115	57.5	8	49.5	1	49.5		
+														
$CuSO_4$	Cu^{++} =62	2	15	332	280	31	249	124.5	15.5	109	1	109		
	S^{+6} =44	1	1	62	36	22	14	7	11	-4	1	-4		
	O^{--} =12	1	17	318	306	6	300	150	3	147	4	588	834.5	
NH_4Cl	N^{+5} =24	1	11	222	200	12	188	94	6	88	1	88		
	H^- = 0	1	23	414	414	0	207		207	207	4	828		
	Cl^- =32	1	7	158	131	16	115	57.5	8	49.5	1	49.5		
+														
$CuSO_4$	Cu^{++} =62	2	15	332	280	31	249	124.5	15.5	109	1	109		
	O^{--} =12	1	17	318	306	6	300	150	3	147	4	588	1662.5	1662

CHAPTER 9

HYDRATIONAL BONDAGE

Incident to the appraisal of the specific gravity of aqueous solutions as an index of hydrational potentialities as reported in Chapter 3 it was noted that in numerous instances a satisfactory order of agreement between calculated and observed values was obtained under conditions in which the solute concentrations were so great as to preclude the existence of solute units with full complements of the prescribed H_2O^- units in hydration. Yet notwithstanding this situation any empirically-assumed sub-masimal distribution of H_2O^- units among the solute ions gave results inconsistent with the observed specific gravity values. Under these circumstances the order of agreement between predicted and observed values for the specific gravity of aqueous solutions of many substances at higher concentrations could be interpreted only as evidence that the prescribed hydrational potentialities had been satisfied by a sharing of the H_2O^- units by the involved solute ions. If this interpretation was correct it appeared to follow inevitably that the shared or jointly held H_2O^- units constituted a medium bonding the involved solute units. This development led to the interesting suggestion that at some concentrations the bonding hydrational H_2O^- units might serve as a flux in the absence of free solvent. In fact, apart from a knowledge of the specific hydrational potentialities of the involved solute ions it might not be feasible or even possible to differentiate a solution in which hydrated ions were distributed within an aqueous solvent and an aqueous liquid in which all of the solute ions possessed full complements of H_2O^- units and no free solvent was present. An analogous situation might hold also for an aqueous liquid in which no free solvent was present and in which there was a sharing of the H_2O^- units by the solute ions. It was obvious that as thus appraised the investigations were leading away from the area of true solutions in which the solute units were to be envisioned as individuals dispersed in a solvent and possessing a measure of freedom which permitted analogy between their behavior and that of gaseous units.

Obviously it was quite by accident that the periodic hydrational potentiality evidenced by the behavior patterns of independent or separate solute units had become involved and recognized as a potential bonding force. Immediately some attractive challenging questions arose. What was the relation of the bonding mediated by the sharing of hydrational H_2O^- units to the chemical bonding mediated by the sharing of electrons? Was the incorporation of integral numbers of H_2O^- units into chemical compounds an expression of the persistence of hydrational bondage? Was there at hand in the description of periodic hydrational potentiality a sound basis for the study of hydrophilic gels and bound water? Was the basic order so beautifully evidenced by crystals in any way subject to correlation with the description of periodic hydrational potentiality? Was hydrational bondage with its involved H_2O^- units subject to appraisal as providing the flux or "oil" for the machinery of protoplasmic metabolism?

There had accumulated through the labors of many investigators a wealth of observational data on the specific gravity of aqueous solutions. These data readily were available in such compilations as handbooks of chemistry and physics, the Smithsonian Tables and the International Critical Tables. These data could have been cited to exhaustion to evidence hydrational bondage, but under the circumstances, with a wide choice, the data for only a few solutions were selected to represent the extensive accumulation. The procedure followed the pattern used in Chapter 3 to validate the description of periodic hydrational potentiality and the cited data comprise Tables 20 and 21.

The data of Tables 20 and 21 included values evidenced by observed specific gravity as having involved and represented completely hydrated solute ions at lower concentrations in which the calculated total volumes of solute present were less than one liter. At these concentrations there was free solvent and the liquid was subject to appraisal as a solution. On the other hand, at the concentrations indicated by heavy line blocking the calculated total volumes of solute present were greater than one liter and no free solvent was indicated. Since the specific gravity values calculated on these bases were of the order of the observed values it was considered to follow that the hydrational potentialities of the involved solute ions had been satisfied by a sharing of the H_2O^- units, a development which conferred a bonding role upon these units. It was of interest that the evidenced sharing of H_2O^- units in hydrational bondage appeared restricted to conditions in which the solvent had been insufficient to permit the degree of solute ion hydration prescribed by the description of periodic hydrational potentiality. Under these conditions with the supply of solvent exhausted the resultant aqueous liquid became subject to appraisal as outside the category of aqueous solutions. The attribute of fluidity in the

Table 20

Data interpreted as evidence of hydrational bondage. Values in thick types represent volumes calculated for independent hydrated ions.

Solute	Ions Wa	Za	Ions Wh	Zh	$\frac{Za}{Zh}$	Conc. $\frac{gms}{liter}$ (Wa)	Conc. $\frac{gms}{liter}$ (Wh)	Solute $\frac{ml}{liter}$ (Vh)	Calc. Sp. Gr.	Obs. Sp.gr.
WO₃	W^{+6} =160	196	W^{+6} =376	1330	.1473	109.6	746	650	1.096	1.096
	O^{--} = 12		O^{--} =318			159.74	1083	946	1.137	1.141
						243.4	1650	**1440**	1.210	1.217
						305.28	2070	**1810**	1.260	1.272
						408.90	2780	**2435**	1.345	1.363
						527.76	3580	**3125**	1.455	1.466
						618.40	4200	**3665**	1.535	1.546
						720.72	4880	**4260**	1.620	1.638
K₂CrO₄	K^{+} = 40	188	K^{+} = 94	566	.332	400.6	1207	906	1.301	1.2921
	CrO_4^{--} =108		CrO_4^{--} =378			451.1	1353	**1015**	1.338	1.3268
						468.5	1411	**1060**	1.351	1.3386
						486.2	1467	**1100**	1.357	1.3505
						504.1	1520	**1140**	1.380	1.3625
						522.3	1572	**1180**	1.392	1.3746
						540.9	1630	**1222**	1.408	1.3858
						559.6	1685	**1263**	1.422	1.3991

Table 21

Data interpreted as evidence of hydrational bondage. Values in thick types represent volumes calculated for independent hydrated ions.

Solute	Ions Wa	Za	Ions Wh	Zh	Za/Zh	Conc. gms/liter (Wa)	Conc. gms/liter (Wh)	Solute ml/liter (Vh)	Calc. Sp. Gr.	Obs. Sp.gr.
AgTl(NO₃)₂	Ag^+ = 96 Tl^+ =164 NO_3^- = 60	380	Ag^+ =474 Tl^+ =344 NO_3^- =348	1514	.252	228.8	908	725	1.183	1.144
						381.6	1514	**1209**	1.305	1.272
						818.5	3250	**2595**	1.655	1.637
						1120.2	4440	**3550**	1.890	1.857
						1323.4	5240	**4180**	2.060	2.036
						1556.8	6170	**4920**	2.250	2.22
						1826.25	7240	**5780**	2.460	2.435
						2144.0	8500	**6780**	2.720	2.680
Cd(NO₃)₂	$CdNO_3^+$=106 NO_3^- = 60	220	$CdNO_3^+$=376 NO_3^-=348	724	.3038	210.3	693	532	1.162	1.1682
						238.1	784	600	1.184	1.1904
						393.7	1300	996	1.294	1.3124
						483.8	1590	**1296**	1.400	1.3822
						583.6	1920	**1472**	1.448	1.4590
						694.7	2288	**1755**	1.533	1.5438
						817.8	2700	**2073**	1.627	1.6356

absence of free aqueous solvent appeared to document for hydration a potential role as a bonding agent in the direction of the solid state.

At this point it became of interest to inquire into the potential significance of the indicated hydrational bondage in relation to "bound water"—a term commonly applied to water held against the stresses imposed by a specific environment. Agar was chosen as the hydrophilic matrix and a number of 5% agar blocks were prepared of uniform size and of a weight approximating 50 gms. These agar blocks were allowed to dry out on a laboratory table until weights of the order of 5 grams had been attained. The blocks then were weighed and immersed in saturated solutions of KCl, NaCl or LiCl, these solutes having been selected to represent different hydrational potentialities. After seven days the blocks were removed and weighed. The increases in weight were compared with the calculated weights of the respective solute ions as derived through the use of the description of periodic hydrational potentiality. The results have been given in Table 22.

From the data of Table 22 it was concluded that the observed increases in weight of the immersed agar blocks had been occasioned by the presence of hydrated solute ions within the blocks and that on a relative basis the evidenced increases in weight were subject to correlation with the hydrational potentialities prescribed by the description of hydration. Similar results had been obtained elsewhere in researches with sections of algal stipe, but the results had been interpreted to signify that the dissolved salts had interfered with the entrance of water into the submerged tissues (see reference at end of chapter).

Following the weighing of the agar blocks which had been removed from the solutions they were left exposed to the air on a laboratory table. Under these conditions the blocks which had been immersed in the KCl solutions gradually became covered on all exposed surfaces with clear needle-like crystals, some of which attained a length of several centimeters. The blocks which had been immersed in the NaCl solutions gradually became covered on all exposed surfaces with white crustose masses of salt. The blocks which had been immersed in the LiCl solutions became reduced to a viscous fluid which remained in that state indefinitely. From the description of periodic hydrational potentiality given in Chapter 2 and repeatedly validated by the data of specific gravity and osmosis the H values for the respective involved solute ions were to be derived as follows: $K^+ = 3H_2O^-$, $Cl^- = 7H_2O^-$, $Na^+ = 11 H_2O^-$ and $Li^+ = 19$

H_2O^-. It was ventured that the hydrational potentiality represented by the Li^+ ion exceeded the hydrating potentiality represented by the laboratory air. If this were true, and potentiality for hydration was an attribute restricted to ions, there was at hand the challenging problem of determining the nature of the atmospheric ions whose hydration was responsible for dehydration in the case of the agar blocks. Appraised in conjunction with the above-indicated H values the ability of table salt to remove moisture from moist air and the inability to retain moisture in dry air suggested for atmospheric ions and sodium or sodium chloride an overlapping range of hydrational potentialities. Once again the researches had directed attention to problems outside the area of solutions, this time in the direction of the gaseous state rather than the solid state.

In researches whose results were reported in the previous chapter it was noted that when copper sulfate was mixed with ammonium or nitrate salts and the aqueous solutions were osmotized, the increases in solution volume were such as to suggest or evidence a complete dissociation of the ammonium or nitrate radicals. This development was not obtained with various other mixtures tried, nor when various other copper salts were substituted for copper sulfate. It was ventured that the cupric ion Cu^{++} alone was not involved in the release of atomic nitrogen and further attention was directed to the behavior patterns of copper sulfate.

The data of Table 3 in Chapter 3 had indicated that in aqueous solutions of the several listed sulfates the sulfate radical SO_4^{--} had remained intact and had hydrated in the precise manner prescribed by the description of periodic hydrational potentiality. It had been noted at the time of the analysis that the specific gravity of aqueous solutions of copper sulfate was not successfully predictable in an analogous manner, but no further attention was given to the matter since the context was concerned with the validation of the description.

Upon a return to the consideration of the specific gravity of aqueous solutions of copper sulfate it was found that the salt was exceptional in that the observational data evidenced that in aqueous solution the sulfate radical became completely dissociated,—on the assumption that it had been intact previously—and that this dissociation had been followed by the hydration of the resultant atomic ions. Yet at the same time the salt appeared to evidence a seemingly unusual degree of accommodation to hydrational bondage through the sharing of H_2O^- units. Data illustrative of these developments have been given in Table 23.

Table 22
Data relating to the uptake of saturated solutions by immersed agar blocks.

Solute	Original Weight in Grams	Final Constant Weight in Grams	Increase in Weight in Grams	Relative Increase in Weight KCl=100	Relative Weight of hydrated Ions KCl=100
KCl	5.9	9.10	3.20	100	100
NaCl	5.79	10.70	4.91	153	152
LiCl	6.10	12.42	6.32	198	203

Table 23
Data relating to hydrational bondage in aqueous solutions of copper sulfate.

Ions	Wa	n	H	Wh	u	Product	Sum	Wa ÷ Wh
Cu++	62	2	15	332	1	332		
S+	44	1	1	62	1	62	1666	.0926
O--	12	1	17	318	4	1272		

Gms. per liter $CuSO_4 \cdot 5H_2O$	Calculated gms anh. $CuSO_4$ per liter (.632)	Calculated Hydrated Ions in gms. per liter (÷.0926)	ml per liter (÷1.926)	Calculated ml Solvent per liter	Specific Gravity Calculated	Specific Gravity Observed
152.9	96.61	1040	952	48	1.088	1.0923
177.0	112.0	1210	1107	- 107	1.103	1.1063
201.7	127.3	1375	1258	- 258	1.117	1.1208
227.1	143.7	1550	1420	- 420	1.130	1.1354
253.0	160	1730	1580	- 580	1.150	1.1501
279.8	177	1912	1750	- 750	1.162	1.1659
307.2	194	2095	1915	- 915	1.180	1.1817
335.4	212	2290	2095	- 1095	1.195	1.1980
364.4	230	2480	2265	- 1265	1.215	1.2146

In Table 23 the upper portion indicates the hydrational potentialities of the listed atomic ions as prescribed by the description of hydration. The extent of the actual departure from the complete satisfaction of these potentialities and the allied sharing of H_2O^- units is represented in terms of solvent deficiency in the blocked portion of the table. As interpreted, within this portion there is no free solvent in the indicated liquid, even though the presence of the shared H_2O^{--} units mediates fluidity.

The evidence for the existence of hydrational bondage associating hydrated solute units in an aqueous fluid devoid of free aqueous solvent presented unexpected problems of reappraisal. In such aqueous fluids, for example, it would be natural to expect a greatly impaired freedom of movement and a consequent reduction in specific ionic conductance under electrical stress. In a subsequent chapter evidence will be cited to indicate that hydrational bondage often is an important factor in chemical reaction.

In the area of physiology hydrational bondage was projected as having supplied a sound basis for the evaluation not only of the cellular behavior patterns involved in the phenomena termed suction tension and turgor, but also of such allied composite patterns as those involved in root pressures and the ascent of sap. The calculated osmotic potentialities of hydrationally bonded glucose, for example, were found to be quite adequate to implement the rise of an aqueous liquid to the tops of the tallest trees. There were new approaches, therefore, to problems of long standing.

LITERATURE CITED

MEYER, B. S., & D. B. ANDERSON (1939). Plant Physiology. D. van Nostrand Co., Inc., N. Y. pp. 112-113.

CHAPTER 10

HYDRATIONAL BONDAGE IN CRYSTALS

In the area of hydration the common incidence of integral numbers of H_2O units in chemical compounds has been so thoroughly documented by chemical analyses as to have become non-controversial. Beyond this point, however, there is much uncertainty. The world of crystals has received a great deal of attention and this attention has revealed the beauty of order, an infinite variety and a challenging complexity. The beauty of order has been evidenced in terms of the symmetry of angles and faces. The infinite variety has been evidenced by snow-flakes. The challenging complexity might just be awaiting correlation and integration with the description of periodic hydrational potentiality.

The primary approach to the study of crystals has been the diagnosis of geometric form. With respect to the general patterning the designs have been differentiated into six classes or types having distinctive characteristics relating to form, face, angle or cleavage. Within these six classes differences and similarities have permitted further characterizations based on composition and attributable to what have been termed the specific habits of the solute ions incident to deposition. At the level of a specific solute a substantial degree of tolerance has been accorded and ascribed to such factors as impurities, vacancies and environment.

Secondarily the study of crystals has involved analyses at the atomic level and the projection or interpretation of arrangements commensurate with gross characteristics, particularly form and cleavage. As a natural outgrowth of the primary approach the emphasis in this phase of study has been placed on the spacial arrangements of the involved atoms. The distinctions between observation and speculation have not been clearly drawn, but in view of the fact that the atomic theory originated in ancient Greece it would not seem appropriate to deprecate the still-superior potentialities of the mind's eye.

The evidenced or projected arrangements of atoms in crystals led to what might be termed the contemporary concept holding the elaboration of macroscopic crystals to be effected by the aggregation of homologous units called building block units. In the implementation of this concept with correlations between crystal designs and observations of the developmental patterns of growing crystals the emphasis continued to be placed on the special disposition of atoms within unit blocks.

From the foregoing considerations it was ventured that the forces bonding the atoms within unit building blocks and the forces involved in the aggregation of unit building blocks to form macroscopic crystals had not received a great deal of attention. It had not been made clear to what extent the internal forces within a unit block were sufficiently balanced : could the unit be characterized as a single molecular crystal? The mechanism of unit block aggregation seemed to need clarification. In the instances in which forces were considered they were appraised as exclusively chemical, even with respect to the development of crystals recognized and designated as hydrated. Here seemed to be a promising frontier of and for research.

In venturing to project inquiry into this area with the description of periodic hydrational potentiality as a background it was recognized that empirical procedures were necessary and that in consequence any attained correlations would emerge in the category of tentative suggestions. With this understanding the procedures followed and the results obtained were assembled to comprise a potential extension involving dynamic aspects of behavior patterns characterizing transitions from solute status to the elaboration of macroscopic crystals.

Arbitrarily it was held that macroscopic crystals were aggregates of single crystals which possessed the following features : (1) a basic structural pattern consonant with the end-products of their aggregation; (2) potential independent status as neutral molecules; and (3) a self-contained potentiality for aggregation which might involve chemical bondage, hydrational bondage, or both. The single crystals were projected as potentially but not necessarily identical with the unit building blocks put forward by other investigators from considerations of the end-products of aggregation. The chemical bonding of single crystals in aggregation was appraised as mediated by the coordinated disruption of multiple bonds and the joint exercise of their potentialities in one or more contrasting directions. The hydrational bonding of single crystals was appraised as mediated by the mutual sharing of the H_2O^- units.

The description of periodic hydrational potentiality as given and validated in preceding chapters formed the basis for appraising the limitations of chemical bondage as well as hydrational bondage. The resulting viewpoint held that atoms could enter into chemical bondage as either positive or negative ions but not as both in covalency. This viewpoint was not that of many chemists who held that in such elements as carbon and silicon covalency existed, and the structural formulas for many carbon-containing compounds commonly were constructed without regard to the maintenance of specific ionic states for the component atoms. The viewpoint also was in contrast to the bondaging potentialities of some molecular solutes whose behavior patterns as noted in a subsequent chapter clearly evidenced balanced amphoteric potentialities for chemical bondage, an allied inability to conduct electricity and a consequent characterization as non-electrolytic. As an intermediate category the solute radicals were appraised as representating a wide range of amphoteric potentialities and differential polarities. In general the gamut of these potentialities extended from marked polarity in inorganic radicals to a polarity in organic radicals which approached the balanced amphoteric potentialities of non-electrolytes.

Since hydrational potentialities comprised a major consideration the differential characterization of solute ions involved restriction to an aqueous solvent. As thus limited the atomic or elemental ions comprised a category distinctly polar and amphoteric only through a modification of electron states in the peripheral orbit. As specific ions they were thus without covalency. As aggregates the solute ions represented a progressive series ranging from dominant polarity to a complete absence of polarity and consequent covalency. The non-electrolytes were appraised as representing a rather distinctive category characterized by collective chemical potentialities for bondage which included the simultaneous operation of positive and negative valencies. The solute radicals were appraised as an intermediate group representing various transitional patterns of potentiality with respect to chemical bondage.

Such data as those given in Chapter 3 had evidenced for numerous solutions at low concentrations a complete hydration of all solute ions present in conformity with the description of periodic hydrational potentiality. These data also had evidenced, as was further indicated in the preceding Chapter 9, that in the absence of free solvent a sharing of hydrational units took place which was progressive with increase in concentration. These developments were interpreted as indicating that within numerous aqueous solutions at the approach of conditions under which crystallization might be expected to take place no solute unit possessed a full complement of hydrational H_2O^- units. Under such circumstances the inclusion of integral numbers of H_2O^- units per solute unit in crystals was projected as representing the involvement of a sustained hydrational bondage.

It was obvious that the major salient in the study of hydrational bondage in crystals had come rather incidentally as a result of chemical analyses and consisted for the most part of gravimetric data evidencing, as noted, the incorporation of whole numbers of H_2O^- units per solute unit. In general the specific numbers of incorporated H_2O^- units bore an inverse relationship to the temperature at which crystalization took place,—a situation which seemed to invite an appraisal of hydrational potentiality as an unquantized force notwithstanding the inclusion of integral numbers of H_2O units and contrary to the evidence given in preceding chapters validating the description of periodic hydrational potentiality. A further disturbing aspect was the fact that in numerous instances hydrated crystals were hygroscopic,—an attribute seeming to testify to an uncertain and frustrating instability. Then again it was obvious that in many cases crystallization could be controlled in such a manner as to result in the formation of either anhydrous or hydrated crystals. Although the culture of crystals had received considerable attention in recent years and a great deal of progress had been made, it was not apparent that any attempts had been made to integrate observed behavior patterns with any description of hydrational potentiality. It seemed appropriate, therefore, to examine the prospects for such an integration.

In the preceding chapter attention was directed to copper sulfate in conjunction with the study of hydrational bondage. The pentahydrate of copper sulfate, $CuSO_2 . 5 H_2O$, as blue vitriol was characterized as crystalline in the triclinic pattern and as non-hygroscopic. At a substantial degree of concentration, as was indicated in the data of Table 23, there was evidence of what appeared to be noteworthy potentialities for hydrational bondage. The pentahydrate therefore appeared to be a suitable substance for inquiry with respect to the relation of hydrational bondage to crystal formation.

The description of periodic hydrational potentiality prescribed for the Cu^{++} ion of weight 62 an ability to take on and hold in hydration 15 H_2O^- units in dilute aqueous solution. This prescribed behavior pattern was validated both by the specific gravity data for aqueous solutions of various copper salts and by the osmotic behavior of the same solutions. Further study indicated further that when Cu^{++} and SO_4^{--} ions were projected as united by a single bond to form a non-polar amphoteric ion $^+Cu - SO_4^-$, the prescribed hydrational potentiality for such an ion, of weight 154, also was 15 H_2O^- units. This meant that the Cu^{++} ion and the $^+Cu - SO_4^-$ ion possessed the same ability to share 15 H_2O^- units in hydrational bondage in the absence of aqueous solvent. It followed that if three sets of copper and sulfate ions were projected as sharing 15 H_2O^- units in the formation of crystals the compositional analysis of the product would be such as to suggest the existence of the pentahydrate $CuSO_4$. 5 H_2O. The three sets also were projected as capable of uniting in chemical bondage to form a single neutral unit crystal. The arrangement and the involved bondages have been indicated in Diagram 1.

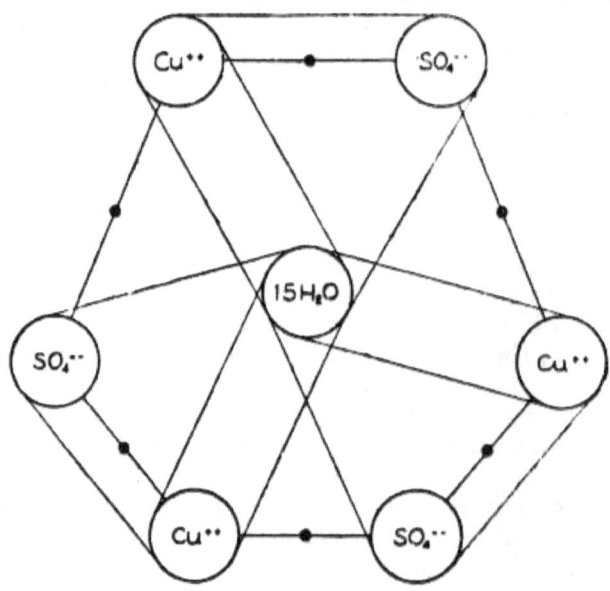

DIAGRAM 1: PROJECTED CONFIGURATION FOR A CRYSTAL
WITH HYDRATIONAL BONDAGE
$CuSO_4 \cdot 5H_2O$

The arrangement given in Diagram 1 projects a configuration involving the cementing of three copper sulfate molecules. The chemical bondages have been indicated as straight lines with central dots representing shared electrons. The curving lines represent hydrational bondages for the Cu^{++} ions and for the Cu^{++} and SO_4^{--} ions in ionic association as $+Cu^+ - SO_4^-$, a state analogous to that of some balanced amphoteric ions to be described and evidenced in Chapter 17. The projected crystal would have a composition which might be expressed as Cu $SO_4.5H_2O$.

Although the proposed arrangement was considered suggestive it was recognized that solid states represented a substantial order of complexity whose resolution scarcely could be anticipated until the specific ion to ion chemical bondages had become quantized. The data of Chapter 6 had indicated that the chemical bonding in the nitrate and ammonium radicals was not subject to simple osmotic disruption, in contrast to the relationship evidenced for some other radicals in the data of Chapter 6. Thus there was the suggestion that nitrogen possessed greater potentialities for chemical bonding, and this suggestion was further substantiated by the osmotic behavior of urea, as may be noted in the data of Chapter 17. In a general way the chemical action and reaction potentialities of atoms and molecules as documented in chemical science serve as practical indices of chemical bonding potentialities, but it would appear that a more precise evaluation must precede the successful prediction of characterizations for many solid state compounds. For example, the observed specific gravity of the copper sulfate represented in Diagram 1 is appreciably greater than that subject to calculation on the basis of the hydrational bondages alone. In Chapter 3 it was evidenced beyond reasonable doubt that the specific gravity of aqueous solutions was a reliable index of the nature and state of the solute. From this premise it would appear inevitable that the specific gravity of substances crystallizing out of aqueous solutions was of analogous potential value as an index of composition. The area is bright with prospects for successful exploration.

It was natural that in the course of an investigation of hydrational bondage in crystals the question arose as to whether or not completely hydrated ions could enter as such into crystals. It had been evidenced repeatedly that hydrated ions could enter into chemical reactions, but no evidence had been attained bearing on their incorporation into chemical compounds in such a state.

In relation to this question the composition of certain crystalline substances seemed suggestive. For example, strontium formate formed crystals in which two H_2O^- units per solute unit were included. As prescribed by the description of periodic hydrational potentiality the formate radical of weight 44 in the anhydrous state would have the ability to take on and hold in hydration one H_2O^- unit. The strontium formate molecule contained two formate radicals. There was the suggestion that the two formate radicals had entered as completely hydrated ions into the formation of unit hydrated crystals. The suggested arrangement has been indicated in

Diagram 2.

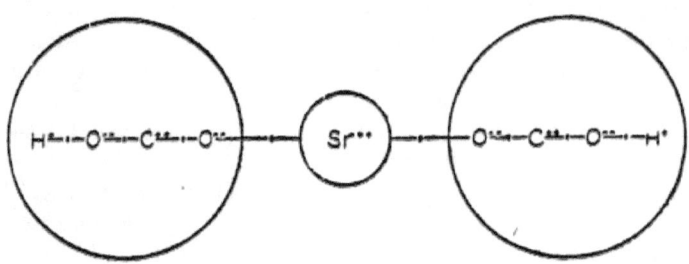

In Diagram 2 the chemical bondages have been represented by straight lines having a central dot. The ionic units projected as having entered into chemical combination in the hydrated state have been encircled,—the indicated relation of hydration bondage projected for copper sulfate in Diagram 1. Up to the present these two types of relationships have been the only ones indicated for hydrated crystals. Either type or both may be represented by specific additional examples.

By far the most exciting aspect of rssearches relating to crystals was the nature of the disclosures inadvertently afforded by further studies of the specific gravity of aqueous solutions. From the data of Chapter 3 there had developed not only a basis for confidence in the significance of the specific gravity of aqueous solutions but also a correlated confidence in the integrity of the composition of the involved solutes as solids preceding solution. Yet in the case of some other aqueous solutions the specific gravity values calculated were not in agreement with observational data which had been obtained by investigators identical or contemporary with those whose values had appeared so excellent in Chapter 3. Thereupon it developed that, based on the specific gravity of their aqueous solutions, some chemical compounds not conventionally represented as hydrated became subject to appraisal as having contained H_2O^- units. Only as thus considered were the calculated and observed values in agreement.

Because these developments in reality were accidental findings they will be taken up in their original context in subsequent chapters. In the general appraisal it seemed to have been made quite clear that hydration was of widespread occurrence and importance in crystal formation.

CHAPTER 11

INTERDIFFUSION

In the two preceding chapters considerations of hydrational bondage have departed from the area of solutions in the direction of the solid state. Other aspects of the study of solutions led in the direction of the gaseous state, and these will be taken up at this point.

In the second chapter of this book attention was called to the theory of Prout enunciated in 1815 and to the fact that the theory had arisen from studies of the behavior patterns of gases. It was noted further that such integral atomic weight values as $O=16$, $N=14$, and $C=12$, values subsequently obvious as twice the atomic numbers, had involved studies of the neutral gases O_2, N_2 and CO_2. At a much later date the atomic weight value $He=4$ was derived for neutral monatomic gaseous helium. In 1866 the Graham law of diffusion was forthcoming. This description apparently met with great favor and had an impressive impact on chemical science.

The theory of Prout had envisioned all of the elements as having been constructed of the smallest or lightest atomic unit, which was the element hydrogen. In the areas of astronomy and geological history contemporary interpretations hold that the incandescent celestial bodies, such as the sun of our immediate system and its analogs in other galaxies, are composed primarily of hydrogen and are the sites of a continuous elaboration of heavier elements from hydrogen. As thus appraised contemporary thought is entirely in accord with the theory of Prout. Within the area of chemistry, however, the theory of Prout long has been deprecated. It was obvious that if the theory of Prout was valid all of the atomic weight values should be subject to representation as whole numbers, multiples of the basic value assigned to hydrogen. Within the area of chemistry it seemed impossible to correlate the irregular fractional atomic weight values with projected integers. Moreover, as mentioned previously, the atomic weight values of chemical science had been derived through researches in which a great deal of painstaking effort had been expended. In some respects the determinations of the atomic weight values for the chemical elements typified the ultimate in meticulous gravimetric procedures. It was natural, perhaps inevitable, that confidence in their integrity was extended to the interpretations involved as represented by the resultant values. This confidence was not impaired when x-ray diffraction patterns led to the assignment of seriate atomic numbers to the elements. It was obvious that such a series of numbers with the element hydrogen as number one conformed to the mathematical prescription of the Prout hypothesis. It was as obvious, moreover, that the conventionally-derived atomic weight values for helium, carbon, nitrogen and oxygen were subject to direct correlation with the assigned atomic numbers. For these elements the atomic weight was twice the atomic number. As noted, these were elements whose atomic weight values had been derived from studies of gases,—as had the Prout hypothesis. In these relationships there was the important suggestion that when elements were in the gaseous state their attributes were more readily and more dependably subject to description and mathematical evaluation than when they were in the solid state. Allied with this suggestion was the prospect that the behavior patterns of solute ions having analogies with gaseous units would prove more amenable to description than elemental units in the solid state. A substantial fulfillment of the prospect has been documented by the data of preceding chapters.

With regard to the historical aspects of the Prout hypothesis as a controversial issue the irregular fractional atomic weight values supplied the principal basis for rejection. As noted, the data of preceding chapters constitute a basis for an alternative interpretation of the gravimetric data involved in the derivation of these irregular fractional atomic weight values and this subject will be taken up in Chapter 18. A secondary but nevertheless important basis for the rejection of the Prout hypothesis was the atomic weight value conventionally assigned to the projected fundamental unit element hydrogen. When an integral unit value of 1.000 was assigned as the atomic weight of hydrogen, the atomic weight values conventionally derived for the elements carbon, nitrogen and oxygen became fractional. Yet the behavior patterns of the gases CO_2, N_2 and O_2 were nicely subject to mathematical description, on which account the derived value $O=16$ attained status as a preferred base for the derivation of atomic weight values for other elements. To the base $O=16$ the derived atomic weight value for hydrogen was 1.008, —a value certainly untenable as a fundamental unit under the Prout hypothesis.

In view of the foregoing considerations the conventional appraisals and apparent behavior patterns of hydrogen were particularly anomalous and disturbing. The element could be studied as a biatomic gas in a state comparable with nitrogen and oxygen. The atomic weights of nitrogen and oxygen were twice their respective atomic numbers, and on an analogous basis the atomic weight of hydrogen should be 2 and the molecular weight of biatomic hydrogen gas should be 4. These values were not in accord with the appraisals of chemical science and this fact prompted a special study of the behavior patterns of hydrogen.

In the first place the data of Table 15 had evidenced quite clearly that the osmotic dissociation of the hydrogen-containing phosphates had released positively charged hydrogen ions of anhydrous weight 4 which possessed the precise potentiality for hydration prescribed for them by the description of periodic hydrational potentiality. To further check on this development the osmotic behavior of some acid sulfates and acid carbonates was studied. The results obtained indicated that an analogous complete dissociation of the radicals took place, followed by a complete hydration of the released element ions including the hydrogen ions. An ionic weight of 4 was evidenced for these positively charged hydrogen ions.

In the second place the data of Table 19 had evidenced quite clearly that incident to the osmotization of specific mixtures there took place a complete dissociation of ammonium NH_4^+ radicals, a release of anhydrous H^- ions of weight zero and a subsequent complete hydration of all ions, including the H^- ions, in precise conformity with the pattern prescribed by the description of periodic hydrational potentiality.

In view of the consistency and degree of precision characterizing the data of Tables 15 and 19 there was no reason to doubt that the H^+ ion had a weight of 4 or that the H^- ion had a weight of zero. These values not only comprised a further validation of the description of periodic hydrational potentiality but also by inference clearly evidenced a molecular weight of 4 for assumed biatomic hydrogen gas. It followed also that as prescribed by Corollary 1 of the description, the weight of a neutral atom of hydrogen would be 2, or twice the atomic number of the element.

In view of these considerations the problem presented by hydrogen was appraised not as a project involving the use of the element in conjunction with a further validation of the description of periodic hydrational potentiality,—a description which already had been repeatedly validated. The problem of hydrogen was that of attempting to explain how it came about in the development of chemical science that a molecular weight of 4 for biatomic hydrogen gas *was not obtained.* From the point of view of the Prout hypothesis the failure to obtain a molecular weight of 4 for biatomic hydrogen gas was tragic, for had this weight been obtained at a time when atoms were appraised as solid balls of the utmost minuteness the conclusion would have been drawn, as was the case with N_2 and O_2, that the atomic weight was one-half the value, or 2. With an atomic weight of 2 for hydrogen the integration of the Prout hypothesis with observational data would have seemed a less formidable project, at least with respect to the lighter elements.

In relation to the molecular weight of biatomic hydrogen commonly one may find that much importance has been attached to evidence afforded by the relative rates at which biatomic hydrogen gas and biatomic oxygen gas have been noted as moving through tubes under specified conditions of temperature and pressure. In such researches various investigators have found that the rate of passage of hydrogen gas was about four times the rate of passage of oxygen gas. There appeared to be no reason for questioning the integrity of the investigators or the results obtained.

With respect to the interpretation of the significance of the data on the relative rates of passage, the involved process commonly was appraised as diffusion. Through the integration of the Avogadro and Graham gas laws the expected relationship became subject to expression as follows when the assumed molecular weight values were $H_2 = 2$ and $O_2 = 32$, and M_1 and M_2 were the respective relative mobilities or diffusion rates.

$$M_1 : M_2 :: \frac{1}{\sqrt{2}} : \frac{1}{\sqrt{32}} .$$

The formula gave irregular fractional values which could be multiplied by 10^4 for convenience to yield for $H_2 = 2$, $M = 7083$ and for $O_2 = 32$, $M = 1768$. As thus calculated and interpreted the diffusion rate for hydrogen gas was about four times the diffusion rate for oxygen. It was obvious that the degree of correlation between the projected and the observed values was excellent and that the dynamic behavior of hydrogen gas as thus appraised was in agreement with gravimetric data interpreted as having evidenced for biatomic hydrogen a molecular weight of 2, or 2.016 to the base $O = 16$. The agreement was considered to be of notable importance and was cited extensively in textbooks of chemistry.

To the writer the foregoing developments appeared worthy of citation also of the potential danger incumbent in zealous concentration. Although the data were obtained under conditions of atmospheric pressure there was no apparent disposition to appraise the movement of the respective hydrogen and oxygen gases as having involved a process other than diffusion. Quite possibly interdiffusion had been interpreted as diffusion. Diffusion was subject to description as a phenomenon attributable to the mutual repulsion of homogeneous units. Interdiffusion, in contrast, was subject to description as a phenomenon attributable to the mutual attraction of heterogeneous units. At atmospheric pressure the observed movement of the respective gases was subject to alternative appraisal as having involved their interdiffusion with air. As indicated previously, the Graham Law of Diffusion was introduced in 1866 and in the subsequent years it had attained prominence in chemistry texts. In contrast, there was *no known law of interdiffusion*. Under these circumstances if data involving interdiffusion were interpreted as data involving diffusion, the error certainly was very natural and excusable. In relation to the matter there has been no disposition to be other than understanding.

In subsequent years desultory attempts were made to adequately describe interdiffusion. In one of these there was projected for the process the formula

$$\frac{m_1 \quad m_2}{m_1 + m_2}$$

in which m_1 and m_2 represented the masses of the interdiffusing units. The description proved to be entirely unsatisfactory whenever light gases were included. Notwithstanding this situation, however, a number of investigations of interdiffusion were carried out. The observational data obtained in these researches included measurements involving such well-established gases and molecular weight values as $N_2 = 28$, $O_2 = 32$ and $CO_2 = 44$, as well as the value 4 for the atomic weight of helium. The inclusion of data involving helium was considered fortunate in relation to a project involving molecular hydrogen prescribed as also of weight 4. Collectively these observational data, presented in the International Critical Tables, were appraised as potentially affording a basis for a recognition and description of the interdiffusion process in mathematical terms analogous to the Graham Law of Diffusion.

For the writer the decipherment of such a description was not easy. There was no pattern for procedure beyond the slow empirical system of trial and error. Yet once successfully ascertained there was nothing but admiration for the meticulous industry of the various investigators who had been responsible for the obtaining of the observational data. Their work seemed all the more noteworthy

for having been carried out in the absence of a prospective or anticipated goal and a consequent opportunity to recheck or repeat experiments whose results seemed divergent.

As empirically evolved by the writer but well integrated with
the observational data of others the description of interdiffusion was
subject to formulation as

$$I = \frac{dm}{dw} k$$

In which I=the coefficient of interdiffusion, dm=the difference
in mobility of the interdiffusing units, dw=the difference in weight
of the interdiffusing units and k=a constant representing a particular
set of standard conditions. A first documentation of this description
of interdiffusion has been given in the data assembled to comprise
Table 24.

In Table 24 the base value for k, used in the prediction of all
subtending interdiffusion coefficients, was derived from the observed
interdiffusion data for O_2 and CO_2. On the assumption that the
utilized description of interdiffusion was correct it was obvious that
the accuracy of prediction was dependent upon the accuracy of the
base value. Since the observational data were obtained at different
times by different investigators the order of agreement between
predicted and observed values was interpreted as attesting the excellent character of the involved research while documenting the
integrity of the involved description· It was to be noted that the
weight values for the interdiffusing gaseous units conformed to the
appraisals of chemical science and that the light gas helium was
included.

It was obvious from the data of Table 24 that the discovery of a
satisfactory formulation describing interdiffusion not only afforded a
hitherto unavailable means of predicting the behavior pattern of
interdiffusing gases but also in a reciprocal manner endowed the
behavior pattern with significance as an index of weight for the
involved interdiffusing gases. Whenever two gases were interdiffusing and the weight of one was beyond question the rate of interdiffusion was conditioned by the weight of the other. Inasmuch as the
molecular weight of biatomic hydrogen was under consideration the
developments made it possible to predict the patterns of interdiffusional behavior for hydrogen gas on each of two assumptions, one
that the molecular weight was 4 as prescribed by the description of
periodic hydration and one that the molecular weight was 2 in practical conformity with the conventional appraisal of chemical science.
These procedures were followed. In the data assembled to comprise

Table 24

Data affording a comparison of observed coefficients of interdiffusion with values predicted from one observed value (top line) through the use of an indicated new description of interdiffusion.

Interdiffusing gases and their mobilities or atomic weights		Relative mobilities of the gases from Graham's law of diffusion		DM/ DW	Predicted Coefficient of Inter- diffusion	Observed Coefficient of Inter- diffusion	Reference for Observed
A	B	A	B				
O_2=32	CO_2=44	1768	1508	21.666	[k=.006263]	.1537	ST
CO=28	CO_2=44	1890	1508	23.875	.1495	.1405	ST
CO=28	O_2=32	1890	1768	30.50	.1910	.1875	ST
CS_2=76	CO_2=44	1147	1508	11.2812	.07065	.063	ICT
CS_2=76	Air=28.8	1147	1865.6	15.2245	.9535	.0892	ICT
N_2=28	O_2=32	1890	1768	30.50	.1910	.1810	ICT
He=4	A=36	5000	1667	104.1562	.652	.6410	ICT

[ST=Smithsonian Tables; ICT=International Critical Tables]

Table 25 the values associated with the conventional appraisal were enclosed in parentheses.

The data of Table 25 were interpreted as having documented the potential value of interdiffusion as an index of molecular weight and as having supplied proof that the molecular weight of biatomic hydrogen was 4 and not 2. The discovery of the law of interdiffusion thereby became evidenced as the enabling act which brought about through data for observed behavior patterns a suitable background for a general recognition of primal basic order at the important cornerstone position in the system of elements. It was considered of interest further that the data had validated within an area of gaseous phenomena a molecular weight value which had been prescribed by corollaries of the description of periodic hydration, a description which had been projected from and pertained to the behavior patterns of solutes.

Following the documentation in the two preceding Tables it was a simple matter to calculate the rates of interdiffusion of hydrogen paired with nitrogen, hydrogen paired with oxygen and of oxygen paired with nitrogen. The description of interdiffusion prescribed zero attraction between homogeneous units, as oxygen paired with oxygen. From the values thus derived the relative interdiffusion rate for hydrogen and oxygen into air became calculable. The data involved and the procedures followed have been indicated in Table 26.

From an examination of the data of Table 26 it was clear that the calculated interdiffusion rate of hydrogen appraised as of molecular weight 4 and air was about four times the calculated interdiffusion rate of oxygen and air. Interpreted as having involved interdiffusion, not diffusion, the observed relative rate of passage of hydrogen and oxygen through tubes at atmospheric pressure constituted evidence that the molecular weight of biatomic hydrogen was 4, as prescribed by the description of periolic hydrational potentiality. Such an interpretation was considered justifiable and correct.

In chemistry the gas laws often have been characterized as the most perfect descriptions in science. These laws have involved homogeneous units and have pertained to the behavior patterns of individual gases as modified by environment. In contrast, the herewith introduced law of interdiffusion pertained to the behavior patterns of associated heterogeneous gases and represented an extension into what could be termed a social area. Implicit in the calculated values was the assumption that the gases involved were in a neutral state and that there was no chemical interaction. The order

Table 25

Data documenting the significance of the pattern of gaseous interdiffusion as an index of molecular weight.

Interdiffusing Gases and their molecular weights		Mobilities of the gases from Graham's Law of Diffusion		$\dfrac{DM}{DW}$	Predicted Coefficient of Interdiffusion k=.006263	Observed Coefficient of Interdiffusion	References* for Observed diffusion Measurements
A	B	A	B				
$H_2=4$	$O_2=32$	5000	1768	115.4285	.72293	.7217	S.T.
$H_2=(2)$	$O_2=32$	(7082)	1768	(177.1333)	(1.10938)		
$H_2=4$	$CO_2=44$	5000	1508	87.30	.54675	.550	I.C.T.
$H_2=(2)$	$CO_2=44$	(7082)	1508	(132.714)	(.83119)		
$H_2=4$	$NO_2=46$	5000	1474	83.95	.526	.535	I.C.T.
$H_2=(2)$	$NO_2=46$	(7082)	1474	(127.4)	(.798)		
$H_2=4$	$CS_2=76$	5000	1147	53.5138	.33515	.3689	I.C.T.
$H_2=(2)$	$CS_2=76$	(7082)	1147	(80.2027)	(.50230)		

*S.T.=Smithsonian Tables.   I.C.T.=International Critical Tables.

Table 26

Data affording a comparison of calculated interdiffusion rates of hydrogen and oxygen into air.

Behavior Pattern	Interdiffusing Gases and Molecular Weights	Mobility Values from the Graham Law of Diffusion	Interfusion as $\dfrac{DM}{DW}$	Average	Ratio
Hydrogen into Air	$H_2=4$	5000			
	$N_2=28$	1890	129.58	122.505	4.01
	$H_2=4$	5000			
	$O_2=32$	1768	115.43		
Oxygen into Air	$O_2=32$	1768			
	$N_2=28$	1890	30.5		

of agreement between calculated and observed values was considered such as to have justified such an assumption. Yet air, composed for the most part of nitrogen and oxygen, commonly has exhibited a capacity for holding moisture, a capacity related to temperature. Did the ability to hold moisture signify the presence of atmospheric ions possessing hydrational potentialities? This was the same question which arose in conjunction with the investigations reported in Chapter 9 and dealing with the dehydrating potentialities of the atmosphere. The question directed attention to the behavior patterns of ionized gases.

CHAPTER 12

BEHAVIOR OF IONIZED GASES

From its inception in the area of solute ions through an extension of the work of Abegg and Bodländer the description of periodic
hydration had involved the principle of change in weight with
ionization. On the $O_2 = 32$ basis this change was evaluated as an
increase of 2 for the accession of each positive charge and a decrease
of 2 for the accession of each negative charge. Within the area of
aqueous solutions the validity of the description of periodic hydration was demonstrated repeatedly, as has been indicated in preceding
chapters. The demonstration, however, involved ions which had
been released from compounds, either incident to solution or to
osmotic dissociation. The behavior of solute ions was such as to
evidence ionic weight values modified from weights of neutral
atoms, such weights having been correlated directly with atomic
number. The inference that change in weight with ionization took
place was substantial, especially since opposite modifications had
been associated with opposite electrical charges and the ionic weights
of the same element with different numbers of charges had been
evidenced. Yet it was to be recognized that the released ions were
subject to appraisal as having been in the same ionic state within the
compound previous to their release, that the existence of a neutral
atomic state represented an inference except for elements of the
helium-radon series, and that many chemists might experience some
difficulty in adjusting to the concept of change in weight with
ionization.

During a survey of the scientific literature on gases conducted in
conjunction with the study of diffusion and interdiffusion it was
noted that researches had been carried out which had yielded
measurements of the rate of movement under electrical stress of
of various gases under conditions of imposed ionization. These data
had considerable appeal and challenge, since from the description of
periodic hydration it seemed inevitable that the ionization of gaseous
units would modify the weight of the units and that by virtue of
such change in weight their mobility would be modified. The observational data were appraised, therefore, as potentially supplying
further proof of change in weight with ionization in the manner
prescribed by the description of periodic hydration.

At the outset of the research here reported it was assumed that the data compiled and presented in the International Critical Tables were adequately representative and reliable. It was assumed further that the experimental procedures had not involved any electrolytic dissociation of molecular gases. In the data cited there were more measurements of ionized hydrogen than of any other gas, and since the molecular weight of neutral biatomic hydrogen gas had been documented as 4 through the discovery of the law of interdiffusion it seemed appropriate to search for a satisfying interpretation of the recorded data for ionized hydrogen gas. It was held, however, that no interpretation would be satisfactory unless it possessed demonstrable facility in the interpretation of the data for other ionized gases.

The compilation of data obtained by various investigators at various times served to supply an important point of reference, on which account the specific values have been copied to comprise Table 27.

Table 27

Data on the mobility of ionized hydrogen gas, as obtained by various workers and assembled in the International Critical Tables.

Mobility Values for	
Positive Ions	Negative Ions
6.70	7.95
6.02	7.68
5.52	8.71
5.34	8.22
5.11	9.67
4.96	8.35

A considerable degree of variation was to be noted in the data. This was not interpreted as evidence of error : the values represented the work of different investigators. It seemed reasonable to hold that different methods were used and that different degrees of success had been attained in the endeavor to impose uniform electrical charges on all of the gaseous molecules.

In attempting to derive a satisfactory appraisal of the data of Table 27 it was considered desirable to seek an arbitrary and tentative point of reference representing the mobility of neutral molecular hydrogen. This point of reference was obtained by adding the twelve values and dividing the sum by twelve. The value thus obtained was very close to the integer 7.0, on which account this integer was adopted as a point of reference.

With 7.0 as a point of reference it became of interest to calculate the relative mobilities of positive and negative ions of biatomic hydrogen on the assumption that there took place incident to ionization the change in weight prescribed by the description of periodic hydration and the further assumption that the Graham Law of Diffusion was operative. Such mobility values might then be compared with observed values. The comparison has been afforded by the data of Table 28.

An examination of the data of Table 28 made it clear that a reasonably satisfactory degree of correlation between calculated and observed mobility values could be attained on the added assumption that in some instances two positive charges had been present on the involved biatomic hydrogen gas molecules.

In the preceding chapter it was made evident that the behavior of gases under atmospheric pressure had involved interdiffusion into air, not simple diffusion. In appraising the data on the mobility of ionized gases there appeared to be no valid reason for assuming that interdiffusion into air had not taken place in an analogous manner. It became of interest, therefore, to examine the observed mobilities in relation to values predicted with the law of inter-diffusion. Data for such a comparison were assembled to comprise Table 29.

An examination of the data of Table 29 made it clear that a reasonably satisfactory degree of correlation between calculated and observed mobility values also could be attained by a similar use of the law of interdiffusion. Both the diffusion law interpretation and the interdiffusion law interpretation evidenced the prescribed change in weight with ionization. It was of secondary interest, but never-theless seemingly important, to endeavor to determine whether dif-fusion or interdiffusion was the primary force which had con-ditioned the mobility of the ionized gases under electrical stress.

The most reliable index of validity for any description in science often had been held to be its usefulness in prediction. In keeping with this viewpoint it was ventured that the involvement of diffusion or interdiffusion as a factor in the behavior of ionized gases might best be revealed by the degree of nicety with which one or the other process demonstrated facility in prediction.

Second only to hydrogen, the behavior of ionized air had been represented by five pairs of measurements for positive and negative ions in the data of the International Critical Tables, and by two measurements for negative ions alone. The reference point 7.0, used in the two preceding Tables, served also as a reference point

Table 28 Data affording a comparison of observed hydrogen mobility values with values predicted with the Graham Law of Diffusion.

Gas and Molecular Weight	Assumed Ionic Wt. Following Indicated Ionization	Mobilities from Graham's Law $\times 10^4$	Mobilities Relative to Neutral Biatomic Hydrogen	Relative Mobility Values $\times 7.0$	Observed Mobility Values
$H^\circ_2=4$	$H^-_2=2$	7082	1.416	9.912	9.67, 8.71, 8.35, 8.22, 7.95, 7.68
		5000	1.000	7.000	7.000
	$H^+_2=6$	4090	0.818	5.726	6.70, 6.02
	$H^{++}_2=8$	3546	0.709	4.953	5.52, 5.34, 5.11, 4.96

Table 29 Data affording a comparison of observed hydrogen mobility values with values calculated with the law of interdiffusion.

Gas and Molecular Weight	Assumed Ionic Wt. Following Indicated Ionization	Mobility from Graham's Law $\times 10^4$	Interdiffusion with N°_2 of Air	Interdiffusion with O°_2 of Air	Interdiffusion with Air as Average of N°_2 and O°_2 Values	Relative Interdiffusion Rate: $H^\circ_2=70$	Observed Mobility Values
$H^\circ_2=4$	$H^-_2=2$	7082	199.7	177.1	188.4	10.76	Negative Ions 9.67, 8.71, 8.35, 8.22, 7.95, 7.68
		5000	129.6	115.4	122.5	7.00	
	$H^+_2=6$	4090	100.0	89.3	94.6	5.40	Positive Ions 6.70, 6.02, 5.52, 5.34, 5.11, 4.96
	$H^{++}_2=8$	3546	77.8	74.1	75.9	4.34	

from which to calculate or predict the mobility values for ionized air.

The data of Table 30 were interpreted as evidence that inter-diffusion and not diffusion had been the process primarily involved in the behavior patterns of hydrogen and air under conditions of imposed ionization. The situation as thus interpreted represented a complexity attributable to the heterogeneity of the ions produced incident to the ionization. The ionization of either biatomic oxygen or nitrogen was evidenced as having yielded less variable mobility values.

The data of Tables 29 and 30 were of interest further in that they evidenced a more extensive positive than negative ionization. Since the description of interdiffusion prescribed no force acting between homogeneous units it was obvious that the calculated values for ionized air, ionized oxygen, and ionized nitrogen would be of the same order of magnitude. On this account the observed values for O_2 and N_2 were included in the table. The average for all negative ions was 1.82, a value which suggested that incident to an imposed negative ionization of air the oxygen alone had been ionized and had become the anion O_2^{--}. The average for all positive ions was 1.36, a value which suggested that incident to an imposed positive ionization of air the oxygen alone had been ionized and had become the sexivalent cation O_2^{+6}. These suggestions seemed to be of particular interest because in terms of valence and charge they reflected the behavior pattern of ionic molecular oxygen. There was also envisioned the possibility that the suggested electron changes might contribute to an improved understanding of respiration.

It was obvious that the law of interdiffusion had evidenced facility in prediction, since from observed values for the behavior patterns of ionized hydrogen gas the behavior patterns of ionized air had been prescribed and observed. On this account it became of interest to survey the behavior patterns of some gases heavier than air under conditions of ionization and electrical stress. In general the observational data for such gases were meager, but at least they could be interpreted as suggestive.

In the tabulated data for ionized gases in the International Critical Tables there was one set of values for argon, and one set of values for hydrogen chloride. From the description of periodic hydration and its corollaries it was obvious that the prescribed weights for these two substances in the neutral state were the same: $A° = 36$ and $HCl° = 36$. Yet the tabulated data for these substances as ionized gases were not of the same order of magnitude. The

Table 30

Data affording a comparison of observed mobility values for ionized air with values calculated with the law of diffusion and the law of interdiffusion.

Gases and Molecular Weights	Assumed Weight of Indicated Ion	Mobility Values Graham's Law $\times 10^4$	Relative Diffusion Mobility	Mobility Values Interdiffusion Law	Relative Interdiffusion Mobility	Observed Values
$H°_2 = 4$		5000	7.000	122.50	7.000	
						Negative Ions
	$N^{--}_2 = 24$	2041	2.86	34.12	1.95	Air : 2.02, 1.87, 1.82, 1.80, 1.80, 1.78, 1.70
$N°_2 = 28$	$N^{--}_2 = 26$	1961	2.75	32.16	1.84	
	$O^{--}_2 = 28$	1890	2.64	32.00	1.82	O_2 : 1.80, 1.79
	$N^+_2 = 30$	1826	2.56	31.00	1.77	N_2 : 1.84, 1.79
	$O^-_2 = 30$					
						Positive Ions
$O°_2 = 32$	$N^{++}_2 = 32$	1768	2.47	30.50	1.74	Air : 1.57, 1.40, 1.37, 1.36, 1.36
	$O^+_2 = 34$	1715	2.40	29.10	1.66	O_2 : 1.36, 1.29
	$O^{++}_2 = 36$	1667	2.33	27.80	1.59	N_2 : 1.27, 1.27, 1.32
	$O^{+6}_2 = 44$	1508	2.11	23.80	1.36	

derivation of a satisfactory explanation of the difference in the behavior pattern became a challenging problem.

With respect to argon it was found that by using the hydrogen base $H°_2 = 7.000$ and assuming that bivalent ions had been involved the calculated or predicted mobility values became $A^{--} = 1.74$ and $A^{++} = 1.40$. The observed values for ionized argon were 1.70 and 1.37 for negative and positive ions respectively. Under the circumstances this order of agreement was considered quite satisfactory: but the calculated or predicted values for HCl were the same as for argon, whereas the observed mobility values for HCl were 0.65 for negative ionization and 0.56 for positive ionization.

With respect to hydrogen chloride it was found that by using the hydrogen base $H°_2 = 7.000$ and assuming that bivalent hydrated ions had been involved the calculated or predicted mobility values became 0.66 and 0.51 for negative and positive ions respectively. Under the circumstances this order of agreement was considered quite satisfactory.

The foregoing developments had introduced the phenomenon of hydration, hitherto considered only in relation to solute ions in an aqueous solvent: into the area of gases. It became of interest, therefore, to search for additional evidence of the hydration of gaseous ions as revealed by the data for the mobility of ionized gases. Using the same reference value, $H°_2 = 7.000$, the observational data appeared to indicate not only that hydration of ionized gases had taken place but also that it had not been controlled incident to ionization. It also became evident that hydration afforded an explanation of the occasional situation in which the observed mobility of the positively-charged ions was greater than that of the negatively-charged ions,—a result which indirectly attested the periodic nature of hydration. Some suggestive comparisons were afforded by data for some heavier gases in Table 31.

Table 31

Data affording a comparison of calculated and observed mobility values for some heavier ionized gases.

		Anhydrous	Hydrated	Average	Observed	
$CO°_2 = 44$	$^-42$	1.35	0.824	1.087	0.81	1.07
	$^+46$	1.26	0.210	0.735	0.73	0.86
$SO°_2 = 64$	$^-62$	0.994	0.242	0.618	0.414	0.415
	$^+66$	0.948	0.265	0.597	0.412	0.414
$Cl_4° = 148$	$^-146$	0.496	0.1865	0.34	0.31	
	$^+150$	0.482	0.182	0.33	0.30	

Table 32

Data indicating the order of expectancy for interdiffusion values following ionization of specified gases.

Gas	Atomic Weight (Neutral)	Mobility×10⁴	Assumed Units Before Ionization		Interdiffusion $H_2 \rightleftarrows Air = 7$			
					+	O	-	Difference
He	4	5000	He°			Base		
Ne	20	2236	He°	Ne°	9.08	9.86	10.80	.94
A	36	1667	He°	A°	5.68	5.94	6.26	.32
Kr	72	1178	He°	Kr°	3.13	3.21	3.29	.08
Xe	108	963	He°	Xe°	2.18	2.22	2.26	.04
Rn	172	763	He°	Rn°	1.43	1.44	1.46	.02

The data of Table 31 were interpreted as constituting further evidence of hydration, since in each of the involved heavier-than-air gases ionization appeared to have been accompanied or followed by hydration in some degree.

Following the survey of the indicated data pertaining to ionized gases it became of interest to make a general appraisal of the specific area of research. Interdiffusion was considered as having been adequately documented as the primary force conditioning mobility. The observed measurements had involved interdiffusion with air, but on a theoretical basis the limitations to be anticipated could be indicated more directly by restricting considerations to the elements of the rare gas series. Some maximum expectancies for data on the ionization of rare gases have been indicated in Table 32.

From the data given in Table 32 it was obvious that even under conditions prescribed as mathematically ideal the differences in mobility attributable to the acquisition of positive or negative electrical charges fell off rapidly as the weight of the gases increased. The data of the two preceding tables had indicated that ideal conditions seldom prevailed, major disturbing factors having been an incomplete and uncertain degree of ionization of the gas and an accompanying incomplete and uncertain hydration of the ions. Under these circumstances the study of the behavior of ionized gases did not appear to offer much promise as an area for future exploitation. The observational data already obtained, however, appeared well worthwhile as documentary evidence relating to the validity of the change in weight with ionization prescribed by the description of periodic hydrational potentiality and also to the validity of the indicated law of interdiffusion.

CHAPTER 13

HYDRATION AND ATMOSPHERIC IONS

The dehydrating potentialities of the atmosphere have been widely recognized as variable geographically and as subject to correlation with temperature and humidity. Yet the existence of dehydrating potentialities within water had been evidenced in data of preceding chapters as having been restricted to an absorptive matrix or solutes having an ionic status. The evidenced hydration of ions attested an ability to abstract H_2O^- units from an aqueous solvent. It could be ventured that the aqueous solvent possessed some ability to retain the involved H_2O^- units against the stresses imposed by the ionized solutes and the question immediately arose as to the nature of the stresses involved in the release of H_2O^- units to air at a water-air interface. The constancy of the relation of temperature to the pressure of water vapor over water suggested that the release of H_2O^- units was primarily a matter of mobility. Yet the relation of temperature to the water-holding capacity of air appeared to have an analogous pattern and raised the question as to whether or not there existed a relationship between atmospheric ions and the ability of the atmosphere to hold moisture. This question was full of challenge. It became of interest, therefore, to undertake studies with a view to the possible identification of atmospheric ions if primarily involved in evaporation and to characterize their patterns of behavior as subject to correlation with the description of periodic hydrational potentiality.

In a general and tentative way some commonly recognized aspects of the atmosphere seemed suggestive. Biatomic nitrogen gas widely was appraised as relatively neutral and inert in comparison with oxygen. If evaporation were to be contemplated as a simple escape or expulsion of H_2O^- units from a water-atmosphere interface the nature of the atmosphere might be somewhat irrelevant, but if evaporation were to be contemplated as involving a correlated hydration of atmospheric ions it appeared unlikely that nitrogen took part in the process. The osmotic data given in Chapters 6 and 7 had indicated that nitrogen linkages were appreciably more tenacious than oxygen linkages insofar as solutes were concerned, and a similar relationship was to be projected for the gaseous state. Under these circumstances specific attention centered upon oxygen as the atmospheric component most likely involved in the event that evaporation from a water-air interface was in any way subject to correlation with the behavior of atmospheric ions.

The data given in Table 30 had indicated that electrical charges could be superimposed upon gaseous nitrogen and gaseous oxygen, but these data involved artificial conditions and had no obvious significance in relation to normal atmosphere. The fact that water saturated with either oxygen or nitrogen gas failed to conduct an electrical current was to be interpreted as evidence that these gases did not undergo dissociation in water. The data given in Chapter 7 indicated that anionic atomic oxygen could be released incident to osmotic activity, but such releases had no obvious direct bearing on the status of atmospheric oxygen.

The interactions of the three salts KCl, NaCl and LiCl with air were interpreted as evidence that atmospheric oxygen was present in a biatomic state. In relatively dry air KCl and NaCl remained dry, but LiCl took up and retained moisture. In moist air KCl remained dry but NaCl took up and retained moisture. Yet moist NaCl would yield its moisture to dry air. It was obvious that NaCl would ionize in water, and the hydrational potentialities of the resultant ions as prescribed by the description of periodic hydrational potentiality were as follows: $Na^+ = 11\ H_2O^-$ units and $Cl^- = 7\ H_2O^-$ units. If biatomic oxygen was projected as subject to simple ionization through collision or otherwise the calculated hydrational potentialities became $O_2^+ = 6\ H_2O^-$ and $O_2^- = 8\ H\ O$. The indicated values for Na^+, Cl^-, O_2^+ and O_2^+ suggested that if atmospheric oxygen played any part in hydrational phenomena it did so in the biatomic state and did not undergo any dissociation.

Three aspects of air appeared to be of special interest in relation to hydration and atmospheric ions. One was the composition of the air in the common environment. One was the maximal amount of water capable of being held by air at different temperatures. One was the vapor pressure of water vapor over water at different temperatures. If observational data for these three aspects were found to be subject to correlation there seemed to be a prospect for associating hydration with the behavior patterns of specific atmospheric ions.

In the common environment the observational data for the composition of dry air indicated that oxygen comprised about 20% and nitrogen about 80%, the relatively small amounts of other gases seeming negligible. On this basis the average molecular weight of dry air became calculable as 28.8, and at 0°C and 760 mm Hg pressure the volume of 28.8 grams of air was calculable as 22.4 liters. At

100°C and the same pressure the calculated volume became 30.6 liters, or the same volume of air would weigh 21.08 grams. The volumes or weights at intermediate temperatures readily were calculable with reference to absolute zero, considered as -273°C.

The maximal amount of water capable of being held by air at different temperatures, 0°C to 100°C, became of interest because if the ability of air to hold water was projected as possibly due to the hydrational activity of atmospheric ions, the amounts capable of being held possessed potential significance in the identification of the nature and status of the involved ions. It was well known that the water-holding capacity of air increased with temperature increase and it followed that if hydrational potentiality were involved, increases in temperature brought about increases in the number of ions present in the atmosphere.

The vapor pressure of water vapor over water was interpreted as a measurement involving air holding a maximum amount of moisture and therefore constituting a measurement paralleling the data on the maximal water holding capacity of air.

Data relating to the approximate content of dry air at different temperatures readily were calculable through the application of the gas laws. Data relating to the maximal waterholding capacity of air at different temperatures and to the vapor pressure of water vapor over water at different temperatures readily were obtainable from handbooks of chemistry and physics. Parallel sets of data relating to the three indicated aspects of hydration and atmospheric ions have been assembled in Table 33.

The data of Table 33 were interpreted as evidence that the moisture-holding capacity of the atmosphere was attributable to the ionization of molecular oxygen. The data of Table 30 had indicated that even under added stress negative ionization of oxygen had been limited to the addition of two charges. The data indicated further that under added stress as many as six electrons had been removed from molecular oxygen. The behavior pattern was suggestive with respect to breathing and respiration, since absorption and osmosis involved added stress, but with respect to the moisture-holding capacity of the atmosphere the data of Table 33 suggested or evidenced that the observational data could be explained as attributable to the modest loss of a single electron by atmospheric molecular oxygen, the moisture held by air representing moisture held through the hydration of O_2^+ ions.

The data of Table 33 were interpreted also as evidence that the vapor pressure of water vapor over water was attributable to the

Table 33

Data pertaining to three aspects of the relation of hydration to atmospheric ions.

TC°	Oxygen* Content of Atmosphere	Relative Oxygen Content of Air	Moisture Held by Air per m³ gms.	Relative Water-Holding Capacity of Air	Vapor Pressure Over Water mm Hg	Relative Vapor Pressure Over Water	Moisture Held per unit O_2 Content of Air, Relative	Vapor Pressure per Unit O_2 Content of Air, Relative
100	4.22	100			760	100		100.000
90	4.34	103			525.76	69		67.000
80	4.46	105.7			355.1	46.6		44.200
70	4.59	108.8			233.7	30.7		28.250
60	4.72	112.0			149.38	19.65		17.520
50	4.87	115.4			92.51	12.18		10.530
40	5.03	119.2			55.324	7.28		6.100
30	5.20	123.2	30.039	4.00	31.824	4.18	3.250	3.390
20	5.37	127.2	17.118	2.25	17.535	2.30	1.770	1.800
10	5.56	132.0	9.330	1.23	9.209	1.21	0.932	0.916
0	5.76	137.0	4.835	0.635	4.579	0.60	0.464	0.436

*Gms. per 22.4 liters at 760 mm Hg pressure.

presence of hydrated O_2^+ ions, and that in the over-water environment all of the O_2^+ ions present at a specified temperature were hydrated. The suggested or evidenced relationships afforded an explanation of the behavior patterns involved in humidity, since as projected the number of O_2^+ ions increased with increase in temperature. The complete hydration of O_2^+ ions in the over-water environment in turn was subject to correlation with a release of the H_2O^- units at the air-water interface. The parallelism between the water-holding capacity of the atmosphere and the vapor pressure of water vapor over water made it seem appropriate to extrapolate the observational data for capacity with the more readily obtainable data for pressure. As projected and evidenced, at the boiling point of water all of the molecular oxygen in the air above water was ionized and hydrated.

The foregoing considerations were of interest in conjunction with an appraisal of atmospheric conditions extraneous to the above-water environment. From the formula for the density of hydrated ions cited in several preceding chapters it was evident that the density of hydrated ions was less than the density of anhydrous ions, a circumstance subject to projection as mediating movements of ions within the sea of inert nitrogen comprising the basic matrix of the atmosphere. The earth's negative charge might well be projected as attracting anhydrous O_2^+ ions and the hydrated O_2^+ ions of lesser density might be projected as rising and as eventually aggregating to form mist and clouds. Speculation may extend the behavior patterns of the anhydrous and hydrated ions to include the concept of the hydrational units as condensers of electrical charge and associate the hydrated ions with electrical phenomena and the precipitation of rain.

The projection of hydration into the area of atmospheric ions is a recent development and unavoidably involves speculation. Another approach is that which concerned evaporational stress in relation to the inclusion of the H_2O^- units in crystals. In general the lower the temperature at which substances crystalize out of aqueous solutions the greater is the tendency to form hydrated crystals. This tendency could be associated with a reduction of the dehydrating potentialities of the atmosphere, such as would result from a projected decrease in the number of atmospheric ions present, or with a decreased hydrational attraction of the solvent for the H_2O^- units, or with both.

It was of interest further that solute units such as CHO^+ and CHO^-, evidenced as having potential roles in protoplasmic meta-

bolism, possessed hydrational potentialities of a magnitude comparable
with those projected for atmospheric ions of oxygen. Theoretically
the presence of such ions within plants appeared to maintain turgor
against the evaporational stresses imposed by the atmosphere,—
stresses apparently related directly to the extent of hydration of the
ions present.

CHAPTER 14

HYDRATION AND THE ABSORPTION OF RADIANT ENERGY

In foregoing chapters the common presence of hydrated ions has been evidenced not only for aqueous solutions but also for solids crystallizing out of aqueous solutions and for gaseous ions above water. These considerations have emphasized the importance of hydration as a universal feature of the world in which living organisms operate and as an endemic feature of all protoplasmic metabolism. Within organisms one interesting aspect of hydration was its potential relation to the absorption of radiant energy. Since the specific heat of water was relatively great it was obvious that its inclusion in protoplasm would facilitate the retention of absorbed radiation. Such a retention has appeared to have an intimate relation to the physiology of land plants, in which the absorption of solar radiation commonly is followed by an increase in temperature above that of the surrounding air. There is a radiation of heat into the air of the intercellular spaces within the leaves, and from the considerations of the preceding chapter it would follow that the number of oxygen ions would be increased by the increase in temperature.

Within the moist intercellular spaces of the leaves the oxygen ions would be expected to become hydrated, and as thus projected the basic accession of oxygen by the mesophyll tissue would involve these hydrated oxygen ions. An analogous situation which might be projected for breathing in warm blooded animals will be discussed in a later chapter. Of major concern at this point was the extent to which hydration modified the absorption characteristics of ions.

Within the area of spectroscopy the emission of radiation attributable to quantized returns of electrons in outer orbits of excited atoms to inner orbits had been differentiated as linear in contrast to the continuous bands of radiation given off by the liquids or solids within an incandescent body. Linear radiation had become diagnostic as an index of excited atoms. Yet in conjunction with a prospective appraisal of the relation of hydration to absorption it became evident that characteristically the conditions involved in the development of the excited states yielding linear radiation were not the conditions affording hydration. In flaming salts or in electrodes, for example, the temperatures were above those commonly present in hydrational phenomena. On the other hand the temperature ranges for hydrational phenomena and for protoplasmic metabolism were about the same. The long-recognized analogy between solutes and gases arose primarily from the relative isolation and freedom of movement of the involved solute and gaseous units. It was recognized that within the temperature range of protoplasmic metabolism excitation and radiation phenomena might take place, as in numerous luminous animals and plants. The radiations thus produced, however, were not of the linear type and hence were outsidethe area of inquiry.

A more promising avenue of approach appeared to be available in the phenomena of absorption. The absorption of radiation by gases was recognized as linear and quantized in the manner of emission. At appropriate levels the same wavelengths characterized emission and absorption. The important question thus became concerned with the extent to which hydration might modify the manner in which ions absorbed radiation.

There was no known source of direct evidence bearing on this question within solutions, although it was recognized that under certain controlled conditions direct evidence eventually might be obtained. Indirect evidence of an appreciable modification of absorption by ions as a result of hydration was available through the color changes which hydration brought about in certain crystals. In copper sulfate, for example, the deep blue color of the pentahydrate is in sharp contrast to the greenish white color of the anhydrous salt, and the blue color persists in aqueous solutions. The data of Table 23 indicated that in aqueous solutions of the pentahydrate all of the solute ions were hydrated.

The potential relation of hydration to the absorption of radiation by ionized gases proved to be a most interesting area of inquiry. The behavior patterns which had characterized the Wilson cloud chamber in use became subject to interpretation as documenting in dramatic fashion the hydrational potentialities of gaseous ions. In the natural atmosphere of the earth these potentialities were assumed to exist, but the exercise of these potentialities was conditioned by the presence of atmospheric moisture insofar as atmospheric ions were concerned. The inquiry led to a study of the so-called Frauenhofer lines, which could be described as shadows on the spectrum of solar radiation attributable to absorption by specific entities between the sun and the observer. On a qualitative basis the wavelengths of radiation specifically absorbed have been indicative of the element involved in the absorption. On a quantitative basis the degree of stability of the absorption has been indicative of the position of the absorbing entities. Lines which decreased from daybreak to noon and increased from noon to sunset have been interpreted as attributable to entities within the earth's atmosphere, since the indicated changes have been correlated with decreasing and increasing depths of the earth's atmosphere through which sunlight has passed during the day.

In the preceding chapter special attention was directed to oxygen appraised as a major source of atmospheric ions. It was quite natural; therefore, that in a study of the Frauenhofer lines special attention was directed to the lines attributed to absorption by atmospheric oxygen. There were two such lines, the A line at 7593.8 Angstrom units and the B line at 6867.2 Angstrom units. These lines varied in intensity during the day and varied in width with changes in the weather. The variation in width could be interpreted as attributable to the presence of moisture and ordinarily there would be no particular need for a more precise analysis. If moisture included the presence of a liquid it was natural that a broadening of the two absorption bands should take place. Conventionally it was projected and held that the A and B Frauenhofer lines were attributable to absorption by oxygen in one or both of two forms, O_4 and H_2O. The explanation seemed sufficiently satisfactory, especially in the absence of any inquiry as to the presence of ionic oxygen in the atmosphere and the relation of hydration to the absorption of solar radiation. Yet the absorption of solar radiation by water takes place in the infra-red region of the spectrum, in very broad seriate bands decreasing in intensity with increase in wavelength. Moisture projected as in a gaseous state might be expected to exhibit a variation in the intensity of its linear absorption commensurate with concentration. Its absorption, however, should remain linear.

In cloudy weather the Frauenhofer lines attributable to oxygen increase not only in intensity but also in width. The development is here interpreted as evidence of the hydration of atmospheric oxygen ions.

Attention at this point may be directed to the results obtained in researches on the light-sensitivity of seeds, as reported in papers listed at the end of this chapter. In these researches it was found that a band of radiation having a maximum at about 7600 Angstrom units was very effective in inhibiting the germination of light-sensitive seeds. It was found also that a narrow band of radiation having a maximum at about 6800 Angstrom units was very effective in promoting the germination of light-sensitive seeds. Under appropriate conditions the inhibiting and promoting effects were evidenced as reversible. The maximum at about 6800 A suggested the involvement of chlorophyll and ether extractions established the presence of chlorophyll within the seeds. The extraction did not contain a pigment having an absorption in the 7600 A region. Very small amounts of radiation were found to be sufficient to incite the respective inhibiting and promoting effects. Moderate inhibition of seed germination was obtained in broad bands of radiation having maxima at approximately 4400 A 4800 A.

The potential significance of the results obtained in the researches on the light-sensitivity of seeds as appraised as a matter of unusual speculative interest. It was obvious that with respect to the major inhibiting and promoting radiations there was a basis for correlation with the A and B Frauenhofer lines. The small amount of radiation necessary to incite a response suggested the involvement of an unstable entity. The reversibility of the inhibitory and promotional effects suggested the involvement of an unstable entity. Each of the critical wavelengths involved suggested that oxygen was the receptive agent. The failure to find a pigment absorbing in the 7600 A region was in keeping with an involvement of oxygen. The association of chlorophyll with the germination response to radiation in the 6800 A region was in keeping with an involvement of oxygen because of the recognized instability present in the nucleus of the chlorophyll molecule and the demonstrable substitution of other elements for the unstable magnesium.

The indicated aspects of correlation between plant behavior patterns and the A and B Frauenhofer lines prompted an examination of the lines attributed to the absorption of solar radiation by hydrogen. There were four such lines : the C line at 6562.8 A, the F line at 4861.4 A, the G' line at 4340.5 A and the h line at 4101.9 A. Three of these lines were of special interest in relation to the sensitivity of plants to visible radiation. The line C at 6562.8 A was subject to correlation with the peak or critical absorption of radiation by chlorophyll B. The F line at 4861.4 A and G' line at 4340.5 A jointly were subject to correlation with absorption peaks of chlorophyll A, chlorophyll B, xanthophyll and carotin, with peaks of phototropic response to spectral light and with the peaks of visible radiation inhibiting seed germination. The correlation of three out of four Frauenhofer lines attributable to absorption by hydrogen with specific aspects of plant sensitivity to radiation made it seem appropriate to survey the lines attributable to absorption by oxygen and hydrogen as a group, and to appraise types of evidence bearing on the significance of the indicated correlations.

The Grotthus Law prescribed that only absorbed radiation was capable of inciting a respose. Although the developments respecting fluorescence have restricted the practical application of this law the correlation of four out of six Frauenhofer lines with absorption peaks of four protoplasmic pigments seemed beyond chance. There was the suggestion that oxygen and hydrogen were intimately involved in the absorption of radiation by plants and hence in the inciting of plant responses. Furthermore, the four involved protoplasmic pigments comprised for all of the higher plants the major physiologically active constituents sensitive to light. It was projected therefore as a general principle that oxygen and hydrogen were not only interceptors of solar radiation when present within the atmosphere but also were interceptors of solar radiation within the aqueous matrix of protoplasm in the chlorophyllous tissues of plants. This principle represented a departure from conventional viewpoints and seemed to invite supplementary considerations.

Two major correlated lines of evidence appear to have contributed to conventional interpretations of the light relations of plants. One of these was the nature of the absorption curves of extracted pigments, commonly separated as chlorophyll a, chlorophyll b, xanthophyll and carotin. The two chlorophylls absorbed throughout the range of the visible spectrum but exhibited trimodal curves with maxima in the general regions 4400 A, 4800 A and 6800 A. The other two pigments absorbed in the range 4000 A to 5600 A, and exhibited bimodal curves with maxima in the general regions 4400 A and 4800 A. The nature of these absorption curves appeared to indicate that the pigments responsible were in either a liquid or a solid state. Correlated with the absorption data were curves obtained in studies of plant responses to spectral light. These curves lacked the nicety of detail which had been obtained with solutions, but in a general way they were subject to correlation with the absorption curves. The data appeared to indicate that the involved pigments were in either a liquid or a solid state.

In a reappraisal of the indicated evidence it was to be recognized that the distinction between chlorophyll a and chlorophyll b involved only the constituents of an H_2O unit, the distinction between xanthophyll and carotin involved only an O_2 unit, that commonly and perhaps universally the pigments were bonded to proteins and that none of the pigments were soluble in water. It was to be recognized further that the peak absorptions varied with the organic solvent and that the range of variation might be as much as 200 A. These considerations did not modify the general nature of the absorption curves, but they did indicate that the critical or peak absorption points were not definite as measured.

It was to be recognized that with respect to the responses of living plants to spectral radiation a re-interpretation of the observational data was long overdue. The impact of commercial fluorescent illumination during the preceding quarter century was such as to emphasize the potentialities of fluorescence in a dramatic manner. Extended to plant physiology fluorescence assumed great potential importance by virtue of the fact that chlorophyll fluoresced in the region of its maximal absorption,—a behavior pattern which seemingly differentiated it from all other pigments. It was obvious that the absorption by chlorophyll of any radiation which induced fluorescence might incite reactions attributable to the induced fluorescent radiation. The differentiation of plant response attributable to fluorescent radiation as compared with plant response attributable to direct radiation emerged as an important challenge to research.

In the immediate area of photosynthesis a distinction between responses to direct and indirect rediation seemed contingent upon the attainment of more information on the minimal and maximal exposures to radiation in the 6800 A region which would sustain the process. Progress in this direction was suggested as dependent upon circumstantial rather than direct evidence. The use of starch as an index of photosynthesis was subject to criticism because starch was a product in the category of a reserve and thus represented a metabolic rate above that adequate for maintenance. Moreover, even in green tissues without starch maintenance often was mediated by the denaturation of organic compounds previously synthesized by photosynthetic activity. For example, in the leaves of bean seedlings grown in complete darkness starch became deposited within the chloroplasts through translocations from the cotyledons. In continuous light the addition of sucrose to the liquid substrate of immersed leaf tissues was found to be followed by a cessation of synthesis and a multiplication of chloroplasts, a result which again emphasized the fact that factors other than light could influence photosynthesis.

An interesting item of potential bearing on the problem was a report that whereas mitochondria were found to be abundant in the mesophyll tissues of specified leaves they were rare or missing in the cells of adjoining palisade tissues. Since mitochondria were to be considered as the structures which mediated protoplasmic metabolic activity there followed the suggestion that the palisade layers of cells served primarily as light screens insuring the incidence of fluorescent radiation in the 6800 A region to the subtending cells of the alveolar mesophyllic parenchyma. Allied with this suggestion was the fact that the number of palisade cell layers bore a direct relation to light intensity : the more intense the light the greater the number of palisade cell layers.

Probably the best evidence of the extraordinary potentiality of radiation in the 6800 A region was to be found in the results obtained with certain small light-sensitive seeds. When such seeds were exposed to spectral light the highest per cent germination and the most rapid and extensive development took place in the 6800 A region. Chlorophyll was found to be universally present in the seeds. When suitable inhibitory radiation was superimposed upon spectral radiation a sharply-linear type of curve for the germination response was obtained at the 6800 A region, a result interpreted as evidence of the involvement of a gas. This 6800 A region, as noted previously, was that of the Frauenhofer B line, and hence in the normal environment would involve a shadow, not a radiation. The energies available in spectral radiation were very low, on which account there was the implication that very small amounts of radiation in the 6800 A region were very effective in inciting a germination response necessarily linked to chlorophyll, since no other pigment was known which had a linear peak absorption in the 6800 A region.

The remarkable efficiency of fluorescent radiation in the 6800 A region in the inducement of seed germination was evidenced in researches in which glass fluorescing this radiation was placed above seeds which were then exposed to spectral ultra violet radiation. Under such conditions very weak lines of ultra violet radiation were included among those which impinged upon the glass and the emitted fluorescent radiation had energy values of the order of one-twentieth of these small amounts. Yet under these conditions seed germination was induced, and a consequent involved activation of chlorophyll was evidenced.

The action involved in the promotion of seed germination by radiation in the 6800 A region was counteracted by radiation in the 7600 A region and the reversibility not only involved short exposures but also could be repeated six or more times. This behavior pattern suggested that a very mobile unit, such as a gaseous unit, was involved. Yet in the natural environment the 7600 A region, as represented by the A Frauenhofer line, was a shadow and analogous to the B Frauenhofer line in that it was attributable to the absorption of solar radiation by oxygen. The common involvement of oxygen in the A and B Frauenhofer lines, the ready reversibility of the activity of radiations at these wave-length regions and the effectiveness of the radiation at very low energy levels comprised correlated evidence that fluorescence possessed potentialities for playing a most important role in the activation of chlorophyll.

The involvement of chlorophyll with photic reactions in seeds appeared to have a great deal of potential significance in relation to photosynthesis. It seemd very improbable that at the indicated critical wavelengths of radiation in the 6800 A and 7600 A regions excitation of chlorophyll would relate to two distinct processes. Thus there followed the suggestion that radiation in these regions had respective promoting and inhibiting effects on photosynthesis. Appraised in this manner the results which had been obtained with light-sensitive seeds inadvertently had afforded a unique avenue of approach to a major problem.

The response of light-sensitive seeds to radiation in the 7600 A region permitted the addition of the A Frauenhofer line to the list of lines previously correlated with plant responses. This meant that five of the six Frauenhofer lines attributable to absorption by oxygen and hydrogen were subject to correlation with the light relations of plants. Although the exception, the h Frauenhofer line, represented a weak absorption, one might nevertheless venture the prediction that eventually a plant reaction to radiation at 4101.9 A would be detected.

Collectively the foregoing considerations led to the viewpoint that in the atmosphere solar radiation was intercepted by oxygen and hydrogen in states involving potential ionization and hydration, yielding the indicated six Frauenhofer lines. It was ventured further that within chlorophyllous plants in the natural environment solar radiation was similarly intercepted, yielding the indicated photic reactions.

Researches in recent years have yielded results affording an impressive degree of correlation between the geological appraisals of the past history of the earth and the biological appraisals of the attributes and transitional adaptations of plants and animals in relation to environment. Considered as among the earliest plants, the blue-green algae evidenced the possession of abilities to thrive at higher temperatures and lower light intensities than the more complex plants which came later. The geological appraisal held that formerly the earth was appreciably warmer than at present and was enveloped by a hydrosphere which greatly reduced the light reaching the earth's surface, especially with respect to the shorter wavelength radiation within the visible spectrum. From the background of such a correlation it has been projected that as the earth cooled and the hydrosphere became impaired through the localization of moisture in the oceans and ice-caps plants and animals progressed slowly towards adaptation to the changing environment. Numerous evidences of the persistence of the impress of the early environment could be cited. Many blue-green algae of the present will tolerate high temperatures, such as those prevailing in hot springs. Many exhibit optimal photosynthetic activity at a light intensity of about 50 F.C., and this order of light intensity also is optimal for human eyes. The light intensity on a present summer day, however, may be of the order of 12,000 F.C.

Two aspects of the photic responses in plants reported in preceding paragraphs were subject to correlation with the geological viewpoint on a broad scale. One of these was the apparent localization of the receptive mechanisms at the positions of reduced solar radiation in the shadows represented by the respective Frauenhofer lines. The other one was the apparent development of facilities for insuring the protective exclusion of the greater portion of solar radiation and for selective screening through fluorescence to provide suitable energy at low light intensities. Both aspects seemed to point to hydrated oxygen ions as having a major role.

REFERENCES

1. FLINT, L. H. (1934). Light-sensitivity in relation to dormancy in lettuce seed. Compt. Red. Assoc. Intern. Ess. Sem. Copenhagen.
2. —(1934). Light in relation to dormancy and germination in lettuce seed. Science 80:38-40.
3. —, & MCALISTER, E. D. (1935). Wave lengths of radiation in the visible spectrum inhibiting the germination of light-sensitive lettuce seed. Smithsonian Miscellaneous Collection 94 (No. 5).
4. —, & MCALISTER, E. D. (1936). The action of radiation of specific wave lengths in relation to the germination of light-sensitive lettuce seed. Compt. Rend. Assoc. Intern. Ess. Sem. Copenhagen.
5. —, & MCALISTER, E. D. (1937). Wave lengths in the visible spectrum promoting the germination of light-sensitive lettuce seed. Smithsonian Miscellaneous Collections 96 (No. 2).
6. —(1938). On the origin of light-sensitivity in seeds. Spectroscopy in Science and Industry. Wiley.
7. —(1938). Immediate problems in the light-sensitivity of plants. Spectroscopy in Science and Industry. Wiley.

CHAPTER 15

HYDRATION AND ACID-ALKALI REACTIVITY

In Chapter 7 the data of osmotic behavior had evidenced the presence of hydrated hydrogen ions in aqueous solution. In Chapters 3 and 9 the specific gravity of aqueous solutions was evidenced as being of potential service as an index of solute nature and status. It was recognized that the context of chapters dealing with the behavior of gaseous ions introduced something of a digression, but the procedure seemed desirable because it was found that interdiffusion appeared to play an important role in relation to acidity and alkalinity.

For many years the attribute designated as acid-alkali reactivity had been interpreted extensively as having been mediated exclusively by H^+ and OH^- ions. The relative abundance of these ions had been assumed to be subject to logarithmic representation and had led to a system of appraisal involving negative exponents and the symbol "pH". At various times seemingly futile objections had been made to the conventional interpretation, and at least one investigator called attention to the inconsistency of holding that all solute units took part in such behavior patterns as boiling point elevation and freezing point depression and in denying all except the H^+ and OH^- ions the privilege of taking part in acid-alkali reactivity. The same investigator noted also that in compliance with the data and law of Kohlrausch the electrical conductance values of solute HCl and NaOH involved sums which included the same values for the Na^+ and Cl^- ions which had been derived from NaCl solutions, and that the assumed presence of either OH^- ions in solutions of HCl or H^+ ions in solutions of NaOH simply was not in accord with measurements made through the use of one of the most sensitive measuring devices known. On some occasions different concentrations of HCl had been used to illustrate "pH", but never at concentrations greater than 1.0 molar, since at this point the supply of exponents had become exhausted. Yet in spite of or notwithstanding these considerations it long had been obvious that the development of various colorimetric and electrometric devices correlated with the "pH" system had done much to foster by implication or induction an acceptance of the conventional viewpoint. It was undeniable that these devices measured something. The question was "what did they measure?" It became a most interesting and challenging

Table 34
Data relating to the specific gravity of some acidic aqueous solutions.

Solute	Ionic Wts W_a	Assumed State	Ionic Wts. As Indicated	Ionic Weight Ratios	Average Weight Ratios	Conc. in gms/liter	Calc. Sp. Gr.	Observed Sp. Gr.	Original State of Solute
HF	H^+ = 4	H	382	.01042	.5052	157.95	1.055	1.053	1 H_2O
	F^- = 16	A	16	1.000					
HCl	H^+ = 4	H	382	.01042	.5052	172.4	1.0765	1.0776	1 H_2O
	Cl^- = 32	A	32	1.000					
HNO_3	H^+ = 4	H	382	.01042	.5052	260.8	1.135	1.134	1 H_2O
	NO_3^- = 60	A	60	1.000					
HBr	H^+ = 4	H	382	.01042	.5052	1148.875	1.762	1.7675	A
	Br^- = 68	A	68	1.000					
HI	H^+ = 4	H	382	.01042	.5052	155.27	1.103	1.1091	A
	I^- = 104	A	104	1.000					
H_2ASO_4	H^+ = 4	H	382	.01042	.5052	178.0	1.1182	1.1128	A
	$HASO_4^-$ = 130	A	130	1.000					
H_3PO_4	H^+ = 4	H	382	.01042	.5052	109.2	1.0725	1.0764	A
	$H_2PO_4^-$ = 96	A	96	1.000					
H_2SO_4	H^+ = 4	A	4	1.000	.670	240.9	1.144	1.147	A
	SO_4^{--} = 92	A	92	1.000					
H_2SeO_4	H^+ = 4	H	382	.0104	.5114	1697.6	2.122	2.120	A
	SeO_4 = 132	A	132	1.000					

question.

In Chapter 9 all of the data on specific gravity evidenced a complete hydration of the involved solute ions. It became of interest to use specific gravity as an index of the nature and status of ions in solutions known to exhibit acidic reactivity. Data assembled in a survey of such solutions have been given in Table 34.

The data of Table 34 were interpreted as having evidenced that in each of the aqueous solutions of listed solutes hydrogen ions were present in an hydrated state with full complements of H_3O^- units, and that in each solution the hydrated H^+ ions were accompanied by anhydrous ions. Since there were notable differences in the acidic reactivity of the solutions it followed that the nature and status of the accompanying ions in some measure conditioned the activity and effectiveness of the hydrogen ions. It was to be noted further that the data afforded no basis for any suggestion that OH^- ions might be present in any of the solutions; to the contrary, the data appeared to exclude such a possibility.

It became of interest to inquire as to the potential significance of the tabulated data in relation to qualitative aspects of acidity. The osmotic behavior of such solutes as Na_2HPO_4 and K_2HPO_4 (Table 15, Chapter 7) had indicated that the released H^+ ions had become hydrated. On the other hand the electrical conductivity of HCl in aqueous solution had been such as to indicate that the H^+ ions had moved in the manner prescribed for the anhydrous state. It seemed apparent, therefore, that in aqueous solution the behavior pattern of H^+ ions was conditioned by the nature of the environmental stresses.

For more than half a century the analogy between solute units and gases had been recognized, but heretofore there had been little or no sound mathematical basis for differentiating anhydrous and hydrated states in solute units. It seemed obvious in the data of Table 34 that the anhydrous ions would have the greater freedom and mobility and on this account would act more like gases than would the associated hydrated units. It became of interest, therefore, to venture an appraisal of the relative potential interdiffusion rates of the listed anhydrous ions, both with respect to their immediate associated units and with respect to air. The description of inter-diffusion given in Chapter 11 was used in this appraisal and the results obtained have been given in Table 35.

The data given in Table 35 appeared to correlate relative potential acidic reactivity with the degree of freedom afforded the projected hydrated H^+ ions by its associates in water and by air. In only

Table 35

Data relating to calculated relative interdiffusion rates of some anhydrous anions with specified associates and with air.

Solute	I⇌Assoc.	I⇌Air	Ratio: I⇌Air / I⇌Assoc.	Notations
HF	5.43	49.6	9.14	Greater than air
HCl	3.59	30.55	8.50	
H_2SO_4	5.90	47.68	8.08	Averages: See Below
HNO_3	2.42	18.45	7.62	
HBr	2.23	16.70	7.50	
H_3PO_4	1.725	12.60	6.96	
H_2SeO_4	1.43	9.64	6.74	
H_2AsO_4	1.45	8.88	6.12	
H_2SO_4 H_a^+	11.37	126.2	11.10	Greater than air
H_b^+	4.50	130.0	2.88	
SO_{4a}^{--}	1.83	3.84	2.10	
Averages	5.90	47.68	8.08	

one solute, sulfuric acid, was the anhydrous H+ ion evideneed as directly involved, and in that situation the associated solute units imposed restrictions upon it. It was to be recognized that the correlations involved the appraisal of solution attributes. Related to external materials affording appropriate absorptive stress on contact one might project an immediate dispersion of the H_2O^- units hydrating the H+ ions and the consequent release and activity of dynamic anhydrous H+ ions.

The foregoing considerations were interpreted as having suggested or indicated that acidic reactivity was associated with the presence of hydrogen ions, commonly in the hydrated state, whenever these ions were present; and that under these conditions the qualitative aspects of acidity were mediated by the degree of restraint afforded by their associated solute ions. There was no evidence of the presence of OH⁻ radicals as associated solute ions.

In extending the investigations it was noted that the specific gravity of some aqueous solutions not readily subject to projection as including hydrogen ions evidenced the presence of anhydrous ions in a manner analogous so the situation indicated in Table 35. Some data for solutions of this type have been given in Table 36.

Table 36
Data for a number of solutes evidencing analogies suggesting potential acidic reactivity.

Solute	Ionic Wts W_a	Assumed Status	Ionic Wts W_h	$\dfrac{W_a}{W_h}$	Average W_a/W_h	Conc. gms./liter	Calc. Sp. Grs.	Obs. Sp. Gr.
NaCl	Na+ = 24	H	222	.108	.554	311.3	1.200	1.1972
	Cl- = 32	A	32	1.000				
KCl	K+ = 40	H	94	.426	.713	226.6	1.1323	1.1328
	Cl- = 32	A	32	1.000				
LiBr	Li+ = 8	H	350	.0228	.5114	107.46	1.0711	1.0746
	Br- = 68	A	68	1.000				
Li_2SO_4	Li+ = 8	H	350	.0228	.5228	342.37	1.225	1.2182
	SO-- = 92	A	92	1.000				
KN_3	K+ = 40	H	94	.426	.713	176.816	1.1028	1.0982
	N3- = 40	A	40	1.000				
$KMnO_4$	K+ = 40	H	94	.416	.713	62.48	1.0365	1.0379
	MnO4 = 112	A	112	1.000				

A comparison of the data given in Tables 34, 35, and 36 suggested that it might be of interest to relate the anhydrous ions of Table 36 to their associates and to air in the manner represented in Table 35. The procedures led to the data given in Table 37.

Table 37

Data relating to the calculated relative freedom of some anhydrous anions in specified associations.

Solute	I⇄Assoc.	I⇄Air	Ratio Air/Assoc.
LiBr	2.4	16.7	6.95
Li_2SO_4	1.96	13.0	6.64
NaCl	5.76	30.5	5.29
KCl	11.88	30.5	2.57
KN_3	10.20	25.4	2.49
$KMnO_4$*	4.78	11.07	2.31

*Positive ion more mobile, even though negative ion was anhydrous

The data of Table 37 were interpreted as representing an extension of Table 35 to include solutes not yielding hydrogen ions in aqueous solution.

- The question arose as to whether or not the "ratio values" given in these two tables represented qualitative aspects of acidity. In all solutes except $KMnO_4$ the anions were indicated as the more mobile under the Graham Law of Diffusion. If this feature were to be contemplated as characteristic of acidic reactivity it seemed reasonable to venture that in solutes yielding cations more mobile than anions the opposite or alkaline reactivity would result. Data pertaining to specific gravity evidencing such a situation have been given in Table 38.

The solutes listed in Table 38 yielded alkaline solutions in water and it became of interest to calculate for the more mobile ions the relative degree of restraint imposed by their associates and by air. The values obtained were arranged in the descending order of freedom, to comprise Table 39.

If the data given in Table 39 represented the relative freedom of the cations with respect to the anions and to air, and this relationship was contemplated as a qualitative aspect of alkalinity, there remained the quantitative aspect. This aspect appeared subject to

Table 38

Data relating to the specific gravity of aqueous solution of some solutes interpreted as yielding cations more mobile than anions, all ions being evidenced as hydrated.

Solute	Wa	Za	Wh	Zh	Za/Zh	Conc. gm/l	Calc. Sp. gr.	Obs. Sp. gr.
NaOH	$Na^+ = 24$	40	$Na^+ = 222$	508	.0788	215.5	1.200	1.1972
	$OH^- = 16$		$OH^- = 286$					
KOH	$K^+ = 40$	56	$K^+ = 94$	380	.1475	52.26	1.0453	1.0452
	$OH^- = 16$		$OH^- = 286$					
Na$_2$CO$_3$	$Na^+ = 24$	104	$Na^+ = 222$	824	.126	160.5	1.142	1.1463
	$CO_3^{--} = 56$		$CO_3^{---} = 380$					
K$_2$CO$_3$	$K^+ = 40$	136	$K^+ = 94$	568	.2394	109.0	1.087	1.0904
	$CO_3^{--} = 56$		$CO_3^{---} = 380$					
Na$_2$B$_4$O$_7$	$Na^+ = 24$	188	$Na^+ = 222$	980	.192	30.822	1.0257	1.0274
	$B_4O_7^{--} = 140$		$B_4O_7^{--} = 536$					
K$_2$CrO$_4$	$K^+ = 40$	188	$K^+ = 94$	566	.333	235.3	1.176	1.1765
	$CrO_4^{--} = 108$		$CrO_4^{--} = 378$					
K$_3$PO$_4$	$K^+ = 40$	208	$K^+ = 94$	1648	.1264	212	1.188	1.1805
	$P^{+5} = 40$		$P^{+5} = 94$					
	$O^{--} = 12$		$O^{--} = 318$					

Table 39

Data relating to the calculated relative freedom of the cations
represented in the data of Table 38

Solute	I $\rightleftharpoons$ Assoc.	I $\rightleftharpoons$ Air	Ratio Air/Assoc.
$Na_2B_4O_7$	0.76	6.18	8.13
K_2CO_3	1.81	12.73	7.04
K_2CrO_4	1.82	12.73	7.00
Na_2CO_3	1.00	6.18	6.18
K_3PO_4	2.10	12.73	6.07
KOH	2.42	12.73	5.27
NaOH	1.25	6.18	4.94

projection as conditioned primarily by solubility in water, and solubility in turn was conditioned by temperature. It became of interest
to modify the ratios given in Table 39 by the molecular solubilities
at 0°C and 100°C. The data comprise Table 40.

Table 40

Data projecting solubility as a practical quantitative aspect of
alkalinity. All solute units evidenced by specific gravity
as hydrated.

Solute	Solubility 0°C	Ratio × Mol.Sol. 0°C	Molar Solubility 100°C	Ratio × Mol.Sol. 100°C	Remarks
NaOH	10.5	52	86.6	910	At 0°C NaOH and KOH positions reversed
KOH	17.38	91.4	31.8	167.5	
K_2CO_3	8.22	57.9	11.48	80.7	
Na_2CO_3	9.682	4.21	6.42	43.8	At 0°C Na_2CO_3 and K_2CrO_4 positions reversed
K_2CrO_4	3.34	23.4	4.21	29.5	
$Na_2B_4O_7$	0.792	6.44	2.80	22.7	

The data of Table 40 were interpreted as evidence that in a practical
way solubility had been a factor in the expression of the attribute

designated as alkalinity. It was recognized that in the data of Table 38 all of the solute units had been evidenced as hydrated, in contrast to the situation represented for the solute units in the data of Table 34. It was ventured, therefore, that in aqueous solutions of electrolytes in which all solute units were hydrated the activity of the solute units was conditioned not only by their relation to each other and to the atmosphere but also by their relation to the solvent. This appraisal was in contrast to that pertaining to solutions involving anhydrous solute units, in which the latter, acting in a manner more analogous to gases, were restrained to a greater extent by their hydrated associates and to a lesser extent by the aqueous solvent.

As interpreted from the foregoing tabulated data the attribute of acidity represented a degree of freedom for cations which was mediated by the presence of anhydrous anions acting like gases and having an interdiffusional relationship to the atmosphere as a sort of potential buoyancy. The attribute of alkalinity represented a degree of freedom for anions which was mediated by the presence of hydrated cations whose relationship to the aqueous solvent primarily conditioned their activity.

In general acid-alkali reactivity was appraised as an attribute of contact for the most part conditioned qualitatively by the degree of freedom accorded solute ions by their environment and quantitatively by solute ion concentration. When environment accorded to cations a domination on contact the activity was evidenced as acidic. When environment accorded to anions a domination on contact the activity was evidenced as alkaline. As thus characterized acid-alkali reactivity was not an attribute mediated exclusively by H^+, and OH^-, nor was it necessarily dependent upon the presence of either ion.

REFERENCE

FLINT, L. H. (1954.) Hydration of solute ions in relation to acidity, alkalinity, and pH. Plant Physiology 9:107-126.

CHAPTER 16

HYDRATION AND CHEMICAL ACTIVITY

Data given in several preceding chapters were recognized as evidencing ionic properties and potentialities differing from activity commonly characterized as chemical. In Chapter 7 there was evidence that osmosis involved stresses in dissociation which seemed more moderate yet more efficient than chemical forces. The sulfate, phosphate and carbonate radicals could be broken down with chemicals, for example, but the action was considered drastic in comparison with the evidenced osmotic separation of the solute radicals into component atomic ions. In Chapter 9 there was evidence that the sharing of H_2O^- units by solute ions in the absence of free solvent could effect a bondage differing from the bondage effected by sharing of electrons which had played so prominent and important a role in chemical science. It was made apparent that hydrational bondage could mediate fluidity in the absence of free solvent and could be maintained in transition into the solid state. In the aggregation of chemical elements into chemical compounds hydrational potentialities appreciably could enhance chemical bondage. In Chapter 8 there was evidence that stresses outside the common chemical category could effect the dissociation of solute radicals impervious to the stresses ordinarily imposed incident to osmosis. There was evidence also of the existence of H^- ions of zero weight, a weight beyond the province of the gravimetric techniques which had played so important a role in the development of chemical science. In the data of preceding chapters, therefore, there has been presented abundant evidence to document both the integrity of the description of periodic hydrational potentiality and its significance in relation to the behavior patterns of atoms. The present chapter will be concerned with differential characterizations of chemical and hydrational potentialities.

The description of periodic hydrational potentiality prescribed ionization as an essential prelude to the attainment of the potentiality and also restricted the potentiality to ions of a specified weight range commensurate with the system of naturally-occurring elements. Quite possibly within chemical science ionization was to be regarded also as prerequisite to chemical activity : the movement of electrons certainly were to be regarded as sufficiently rapid to defy detection. The dissociative powers were well known and by virtue of these powers moisture alone could effect ionization as a prelude to chemical reaction. With less assurance hydrated ions could be considered as potential catalyzers of chemical activity. In general it seemed to be quite safe to assume that ionization was a prerequisite to both chemical activity and to hydration.

There was no obvious essential relationship between the potentialities for chemical reactivity and hydration, though often they were allied. From the description of periodic hydrational potentiatity it was axiomatic that hydrated ions weighed more and consequently moved more slowly than the same ions in the anhydrous state when both were free to move in the same matrix. Yet it was clear also that in the absence of free aqueous solvent the existence of shared H_2O-units or hydrational bondage,—as might be projected for protoplasm—would impose restrictions upon the movements of hydrated ions. In general, therefore, hydrational potentiality coupled with the appropriate H_2O^- units was subject to interpretation as modulating the tempo and completeness of chemical reactivity. It had long been recognized that within protoplasm chemical reactions took place which were difficult or seemingly impossible to induce in vitro. Any attempt to analyze the complexities of protoplasmic metabolism at this time seemed quite premature, but nevertheless within the context of preceding chapters there was the basis of a hope for eventual success in that area.

The evidence for the persistence of hydrational potentiality into the gaseous and solid states appeared sufficient to permit the venture that chemical bondage and hydrational bondage both covered the entire gamut of atomic experience. As thus appraised it became of interest to compare the periodicities of chemical behavior with the periodicities of hydrational behavior. Data providing for such a comparison have been given in Graphs 1 and 2.

It was to be noted in Graph 1 that the periodicity after Mendeleev, which involved physical attributes and chemical behavior pattern, comprised five complete periods and two incomplete periods. An incomplete and abbreviated initial period was followed by two short and regular periods. These were followed by three long complete but irregular periods and the terminal period was short and incomplete. The specific physical attributes of the elements were appraised as dependent upon composition. The specific chemical behavior patterns were appraised as dependent upon the disposition of the electrons in the peripheral orbits following common types of ionization. Historically the periodicity after MENDELEEV although ridiculed at its inception proved invaluable through its facility in

141

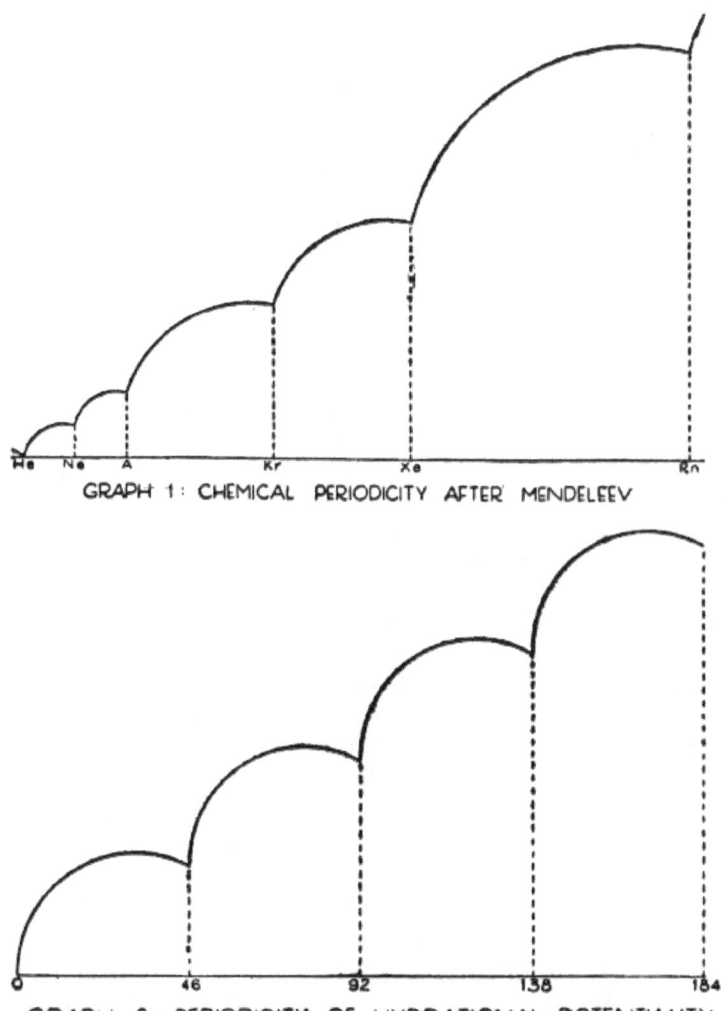

GRAPH 1 : CHEMICAL PERIODICITY AFTER MENDELEEV

GRAPH 2 : PERIODICITY OF HYDRATIONAL POTENTIALITY

prediction and for many years in charted form it has been the outstanding academic show-piece of chemical science.

It was to be noted in Graph 2 that the periodicity involving the
hydrational potentialities of ions comprised four equal regular and
complete periods collectively encompassing the weight range O-184,
commensurate with the ninety-two naturally-occurring elements.
The nicety of correlation carried the implication that moisture had

been involved in the synthesis of the elements. The hydrational potentiality was evidenced as exclusively a property of ionic weight, and within the prescribed weight range was equally characteristic of atomic and molecular ions. This situation was interpreted as evidence that hydrational potentiality primarily involved the aggregate protons and hence was a more fundamental feature than the allied pattern of distribution of electrons in the peripheral orbit. The regularity, completeness and range of the repeatedly evidenced periodic hydrational potentiality was interpreted as consistent with such an appraisal.

It was inevitable that curiosity regarding the anatomical aspect of hydration should develop. In the literature of the preceding century regarding hydration water molecules were pictured as spheres clinging to the periphery of ions, where they were considered as held with sufficient force to evidence collective transfer under electrical stress. Yet to thus represent hydration in conjunction with the description of periodic hydrational potentiality appeared inconsistent with the description itself: its integral nature, its reciprocal nature, its seriate nature, its periodic nature. On the other hand, to picture hydration as involving a varying internal capacitance for the H_2O^- units was not entirely devoid of complications. From the law of Avogadro it was evident that all gaseous atoms and molecules were of the same size under the same set of environmental conditions. The analogy between gaseous units and solute units, even hydrated solute units, appeared to extend to volume, and the specific gravity data given in Chapter 3 evidenced solute density to be conditioned indirectly by the atmosphere, directly by hydrostatic pressure. The density formula for hydrated solute ions carried an implication that hydration decreased density either by increasing volume or by introducing buoyancy. The fact that hydrated ions had mobilities in keeping with the Graham Law of Diffusion carried the implication that volumes were equivalent. There were the resultant suggestions that hydration involved the inclusion of H_2O^- units within the orbital periphery of the anhydrous units, that the included H_2O^- units added buoyancy and weight but did not impair the elasticity which characterized gases and conditioned the behavior documented in Boyle's Law. As thus projected hydrated ions appeared to represent entities intermediate between the gaseous and liquid states. In true solution their behavior patterns were those of gases. In the absence of solvent they exhibited attributes of liquids.

The anatomical features of hydrated ions unavoidably were largely inferential and speculative. The extent to which the behavior patterns of anhydrous and hydrated ions could be differentiated became a matter of special interest and inquiry. The data of Chapter 15 had evidenced that anhydrous ions had important potential roles in acidic reactivity. The behavior patterns described in Chapter 4 had evidenced that absorptive membranes were permeable to anhydrous ions and impermeable to their hydrational complements. Contrary to the implications of the Law of Kohlrausch it had been suggested that a specific ion might be anhydrous in some associations and hydrated in other associations. Yet it was held that what seemed like idiosyncrasies must yield eventually to recognition as behavior patterns motivated by mathematical forces.

The data of Chapter 7 had indicated that osmosis had potentialities for disrupting the chemical linkages in some solute radicals. From a chemical standpoint it became of interest that incident to these disruptions there took place a complete hydration of the resultant ions. These hydrations by virtue of the transfer of H_2O^- units from solvent to solute had a marked effect on the concentrations of the solute. It was natural that the chemical reactant potentialities of the dissociated solute radicals were different from those of the intact solute radicals. Advantage was taken of this situation to make the reconstruction of radicals following their complete dissociation and the subsequent hydration of their component atomic ions, as evidenced by precipitation reactions in appropriate mixtures. Mixtures of non-osmotized solutions of $Ca(NO_3)_2$ and K_2HPO_4 yielded precipitate. Mixtures of osmotized solutions of these substances were clear, but yielded precipitate on the removal of water through heating or through the addition of absolute alcohol. On standing in an open beaker such mixtures through an evaporational loss of water became coated over with an icing of insoluble phosphate. Analogous developments were obtained with mixtures involving sulfates and carbonates. From these results the conclusion was drawn that the reconstruction of the radicals took place in the absence of sufficient aqueous solvent to effect the isolation of the hydrated atomic ions. There thus was the suggestion that hydrational bondage had potentialities for facilitating chemical reactions and chemical bondage.

With these suggestions regarding what might be interpreted as the synthesizing potentialities of concentration it became of interest to consider the evaporational residues obtained in experiments described in Chapter 10, and to use the procedures in further researches. In the evaporational residues of very dilute solutions it was obvious that filmoid depositions were prominent, and in general these exhibited vacancy reticulations with rounded edges. The filmoid nature of the residues was interpreted as attributable to a predominating influence of adhesive, cohesive and hydrational forces, in the order listed. The crystalline nature of the residues of more concentrated solutions, on the other hand, was interpreted as attributable to a predominating influence of hydrational, cohesive and adhesive forces, in the order listed. If these interpretations had merit it seemed to follow that it should be possible to obtain some degree of differentiation of anhydrous and hydrated ions by means of evaporational residues.

Without exception all of the solute units in osmotized solutions were hydrated. It was obvious that whereas under the conditions of laboratory research in which the total increases in solution volume were sought for potential correlation with the description of periodic hydrational potentiality it was essential that an extraneous supply of solvent sufficient to permit complete hydration for all solute units involved, in nature the extraneous supply of solvent often would be limited for many land-based organisms. The situation had interesting speculative aspects involving the scarcity of precipitates in vacuoles and the maintenance of turgor in vacuolar tissues.

From the viewpoint of chemical activity the respiratory patterns of some plant tissues immersed in water and in aqueous solutions assumed to supply analogs of oxygen proved to be of considerable interest. In replicated tests with resting buds of sugarcane it was found that at 25°C the buds could withstand immersion for only about 8 days before showing signs of injury upon removal to moist chambers for germination. Buds immersed in 0.01 molar solutions of Na_2SO_4 could withstand immersion for about 12 days before injury, but required a longer period for germination. Buds immersed in a 0.01 molar solution of Na_2SeO_4 could withstand immersion for about 18 days without obvious injury except that they required an even longer period for germination. Buds immersed in a 0.01 molar solution of Na_2TeO_4 in some instances withstood immersion for a period of 28 days, but required a still longer period for germination. The germination periods suggested that to some extent the chemical analogs of oxygen had taken part in respiratory metabolism. The results were interpreted as consistent with the data given in Tables 15 and 16 in which an osmotic dissociation of sulfate, selenate and tellurate radicals had been evidenced.

Some respiratory vapors exuded by the sugarcane buds during and after immersion were as follows: from water, H_2O; from the

Na_2SO_4 solution, H_2S, from the Na_2SeO_4 solution, H_2Se; from the Na_2TeO_4 solution, H_2Te. The release of these vapors was interpreted as further evidence that the chemical analogs of oxygen had taken part in respiration. There was the suggestion, also, that in water the dissociation of an analogous radical, OO_4^{--}, had been involved. The transitions from 6- valent positive ionic states to valent 2- valent negative ionic states evidenced in the immersion experiments were obscure, yet involved changes in hydrational potentialities as well as in electron redistribution. It was ventured that in many instances the interplay of chemical and hydrational potentialities, presently confusing, eventually would be resolved with the attainment of more precise evaluations of the respective forces.

CHAPTER 17

SOME BEHAVIOR PATTERNS OF REPRESENTATIVE ORGANIC SOLUTES

In preceding chapters particular attention has been directed to element ions and radicals subject to characterization as inorganic. The basic feature of the considerations was the description of periodic hydrational potentiality and this description limited the potentiality to ions. In the area of aqueous solutions the solutes uniformly were subject to characterization as electrolytes. It was natural that the investigations were extended into the area of solute non-electrolytes, particularly since in a general way the hydrational potentialities of some solute sugars long had been recognized.

During the course of researches described in preceding chapters it had become evident that hydrational potentiality was an ionic attribute which characteristically found fullest expression in osmosis but which could in whole or in part be correlated with specific gravity and also find expression in such a manner as hydrational bondage to bring about the inclusion of H_2O^- units in specified acids and crystals. These considerations afforded a pleasant and promising approach to the study of the behavior patterns of non-electrolytic solutes.

In entering upon a new area of research there was natural uncertainty as to how much reliance safely could be placed on contemporary viewpoints. It had become very evident that element ions entering into chemical compounds maintained an ionic integrity; but such an integrity, especially in the case of carbon, had not been honored universally in the construction of structural formulas for numerous organic compounds. Thus whereas some chemists decried covalency for element ions others subscribed to it and innumerable structural formulas for organic compounds ignored it.

With respect to solute radicals it seemed to be quite clear that among compounds of an organic nature amphoteric potentialities existed as an accompaniment of polarity. One could venture, therefore, that non-polar amphoteric ions could exist, could possess the hydrational potentialities prescribed by the fundamental description and yet could not take part in the conduction of an electrical current because of the presence of positive and negative charges on the same ions. The procedures of necessity were empirical; it was inevitable that interpretations would be controversial. Many obvious failures were encountered; the researches reported were restricted to those appraised as satisfying because of suggested significance and potential value. Various types of behavior patterns were evidenced, on which account it seemed advisable to select individual compounds as representative.

At an early stage of the investigations it became clear that since carbon exhibited four common ionic forms, C^{+2}, C^{-2}, C^{+4} and C^{-4}, theoretically any non-polar carbon compound which contained the bivalent cation could have an analog in which the carbon was a bivalent anion; any carbon compound which contained the tetravalent cation could have an analog in whch the carbon was a tetravalent anion. In conjunction with this development it was apparent that the viewpoint holding that the component atoms of a compound maintained an ionic status which demanded recognition in structural formulas was a viewpoint which held special promise in the area of polyatomic non-electrolytic solutes.

A STUDY OF SUCROSE

Sucrose as the most widely known and extensively used carbohydrate became a subject of primary interest, and as the investigations continued it became increasingly obvious that the behavior patterns of sucrose could be considered as representative of the behavior patterns of solute non-electrolytic carbohydrates in general or as a group. The conventional contemporary appraisal of sucrose assigned to it a compositional formula of $C_{12}H_{22}O_{11}$ and a molecular weight of 342.176. These values were incompatible under the description of periodic hydrational potentiality which prescribed a weight of 2 for the neutral hydrogen atom. As previously noted it had not been a common practice in chemical science to accord specific ionic status to the atomic constituents in the projection of structural formulas. Data in preceding chapters had emphasized the importance of such assignments and on this account any appraisal or projection of a structural formula for sucrose inevitably was to become empirical and controversial.

It was of interest to find that within the area of chemical science the structural formula for sucrose had been controversial for some years. Some chemists preferred to represent the sucrose molecule as two oxygen-linked chains while other chemists preferred to represent it as two oxygen-linked rings. The two linked components however represented were appraised as glucose and fructose, or as their respective synonyms dextrose and levulose. The basic problem in the researches on sucrose was the problem of integrating the observed behavior patterns of sucrose with the precepts concerning hydrational potentialities which repeatedly had been validated by the data of preceding chapters.

Hydrational potentiality had become evidenced as an attribute not only restricted to ions, but restricted to ions within a specified weight range. The solubility of sucrose in water implied that hydrated ions within the specified weight range were present, and on this account it was assumed that in aqueous solution a separation of the two component units took place to yield ions within the specified range in weight. In keeping with previous considerations it was projected that two types of sucrose molecules might exist, one in which the two component groups were bonded by a bivalent oxygen anion and one in which the bonding oxygen atom was a bivalent cation. As thus projected one type was the mirror image of the other, the two types differing only in the nature of the electrical charges on the atoms in the respective positions. The two projected types have been indicated in Diagram 3.

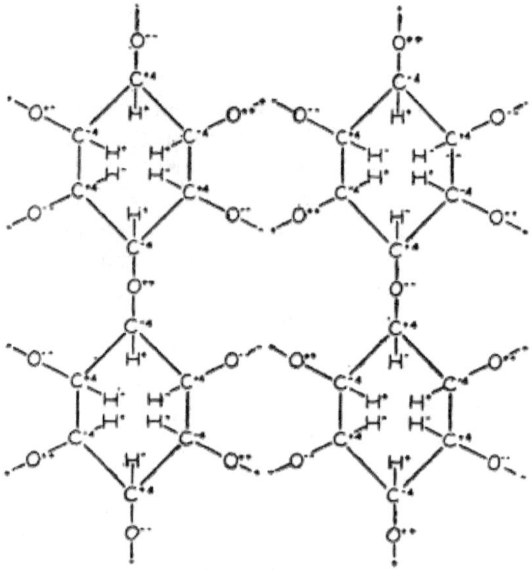

DIAGRAM 3 PROJECTED CONFIGURATION FOR IONIC COMPONENTS OF SUCROSE

As represented in Diagram 3 the molecules were ions capable of being alternately superimposed to form hexagonal doubly-terminated crystals twice as long as broad. These crystals, however, were projected as capable of dissolving in water by virtue of separations at the oxygen linkages and a subsequent hydration of the resultant ions to form isomers. Characterizations for these projected solute units have been indicated in Table 41.

Table 41

Data affording characterizations for four projected sub-units of sucrose.

Ions	Wa	n	H	Wh	Vh	Characterization
$C_6H_6O_6{}^{+3-3}$	180	4	2	216	118	Amphoteric, non-polar glucose isomer
$C_6H_6O_6{}^{-3+3}$	180	4	2	216	118	Amphoteric, non-polar glucose isomer
$C_6H_6O_5{}^{+4-2}$	168	4	8	312	203	Amphoteric, sub-polar fructose isomer
$C_6H_6O_5{}^{-4+2}$	160	4	12	376	264	Amphoteric, sub-polar fructose isomer

In Chapter 10 it had been indicated that molecular ions might enter into chemical compounds in completely hydrated states, the numbers of H_2O^- units involved in complete hydration having been those prescribed for ions of their weight by the description of periodic hydrational potentiality. From the data of Table 41 it was a simple matter to calculate the specific gravity of a mixture of the projected hydrated ions with the density formula given previously. This calculated value was 1.614, but the observed specific gravity of crystalline sucrose as reported in various reference tables was 1.587. The observed value suggested that in the manufacture of sucrose the indicated ions had retained one additonal H_2O^- unit in hydrational bondage. On such a basis the calculated and observed values would be in agreement and the crystalline product would be subject to appraisal as 96.8% pure in terms of the projected hydrated ions. Using this value the specific gravity of a series of aqueous solutions of sucrose over a representative range of concentrations was calculated. The results obtained then were compared with observational data as given in numerous tables. The procedures followed and the results obtained have been indicated in Table 42.

Table 42. Data affording a comparison of calculated and observed specific gravity values for aqueous solutions of sucrose.

% Sucrose	Gms per liter	Calculated Gms Anhydrous (×.968)	Calculated ml solute per liter (÷ 1.614)	Calculated ml Solvent	Calculated Sp. Gravity	Observed Sp. Gravity
10	103.8	100.6	62.3	937.7	1.0383	1.0381
20	216.2	208.5	129	871	1.0795	1.0810
30	338.1	327.3	203	797	1.1243	1.1270
40	470.6	456	282	718	1.174	1.1764
50	614.8	596	369	631	1.227	1.2296
60	771.9	748	463	537	1.285	1.2865
70	943	912	564	435	1.347	1.3472
80	1129	1093	678	322	1.415	1.4117
89	1311	1270	786	214	1.484	1.4732

Table 43. Results obtained in a study of the osmotic behavior of sucrose in aqueous solution

Grams Sucrose	Volume of Aqueous Solution ml	Volume at end of Osmosis ml	Calculated Original Molecular Concentration	Observed Volumetric Increase ml	Standardized Observed Volumetric Increase ml	Remarks
3.44	100	110	.10	10	1000	Required 7 days
6.88	100	120	.20	20	1000	Required 7 days
10.32	100	130	.30	30	1000	Required 7 days
13.76	100	140	.40	40	1000	Required 7 days
17.20	100	150	.50	50	1000	Required 7 days

In Table 42 it was made apparent that the order of agreement between calculated and observed specific gravity values was such as to indicate that the projected assumptions at least were compatible with the involved observational data taken from conventional tables.

At this point it was held that if tho foregoing interpretation of solute sucrose was tenable there was at hand a procedure for calculating the osmotic behaviour of sucrose in conjunction with the description of periodic hydrational potentiality. It was well known that aqueous solutions of sucrose possessed osmotic potentialities, but insofar as known the relation of these potentialities to hydration had never been ascertained. It became of interest to study the osmotic behaviour of sucrose in aqueous solution. The results obtained have been given Table 43.

It was clear from the results given in Table 43 that on a standardized basis the osmotic behavior of sucrose in aqueous solution was uniform. The next problem became one of deriving a satisfying interpretation of the significance of this uniform behavior. The data of Table 42 had suggested or evidenced the presence in solution of the ions assumed as indicated in Table 41. The expected osmotic increase in solution volume attributable to the activity of such ions was calculated in accordance with the procedures followed in Chapter 5, 6, 7 and 8. When this was done it became obvious, as noted, that little or no osmotic activity could be expected from the assumed ions. These calculations have been indicated in the upper portion of Table 44.

Fortunately, the discovery of osmotic potentialities for dissociation as reported in Chapter 7 supplied a basis for assumptions and procedures. It was ventured that osmosis effected a disruption of the carbon-to-carbon linkages in the solute units given in Diagram 3, and on the basis of this assumption the osmotic behavior patterns of the resultant ions were calculated. When this was done, the order of agreement between calculated and observed osmotic behavior patterns was found to be well within the range of unavoidable error in measurement. These developments have been indicated in the lower portion of Table 44.

From the data of Tables 41 to 44 it was ventured that the sucrose which had been involved was a crystalline hydrate having the general composition $2 C_{12}H_{12}O_{11} . 1 H_2O$, that in aqueous solution it formed amphoteric ions and that osmosis brought about a disruption of carbon-to-carbon linkages which was followed by a hydration of the ions which had been thus released.

The results obtained were appraised as of interest and potential

Table 44
Data affording a comparison between calculated and observed osmotic behavior patterns for sucrose.

Assumed Original Solute Ions	Wa	n	H	Wh	Vh	Va	D_1	$\dfrac{D_1}{2}$	$\dfrac{Va}{2}$	D_2	u	V_i	Vt	O
$C_6H_6O_6{}^{+3-3}$	180	4	2	216	118	90	28	14	45	-31	2	-62		
$C_6H_6O_5{}^{+2-4}$	160	4	12	376	264	80	184	92	40	52	1	52		
$C_6H_6O_5{}^{-2+4}$	168	4	8	312	203	84	119	59.5	42	9	1	9	-1	1000
Assumed Disruption of Carbon to Carbon Linkages														
COH^{-1+2}	32	1	7	158	131	16	115	57.5	8	49.5	$\dfrac{10}{2}$	247.5		
COH^{+1-2}	28	1	9	190	166	14	152	76	7	69	$\dfrac{10}{2}$	345		
CH^{-3}	8	1	19	350	342	4	338	169	2	167	1	167		
CH^{+3}	20	1	13	254	236	10	226	113	5	108	1	108		
O^{++}	20	1	13	254	236	10	226	113	5	108	$\dfrac{1}{1}$	54		
O^{--}	12	1	17	318	306	6	300	150	3	147	$\dfrac{1}{2}$	73.5	995	1000

importance, since it seemed to follow that the released amphoteric COH radicals might represent the units involved in translocation within plants and animals and might also represent a common intermediate stage in the synthesis as well as in the breakdown of more complex organic compounds. There was the allied ionic atomic oxygen subject to projection as of dynamic potency in metabolic processes. In general there was the suggestion that osmotic behavior held much promise as a tool in the extension of research in the organic area.

The results obtained also were appraised as frustrating. They advanced again the question which had been raised earlier with respect to the nitrate and ammonium radicals. Did the suggested or evidenced COH units released through the osmotic disruption of carbon linkages represent end-points in denaturation, or was the complete breakdown of these units a prerequisite before utilization of the involved atoms in synthetic processes? In the case of the NH^+_4 and NO_3^- radicals there was no recognized incorporation of the intact units in organic compounds, but the situation was different with respect to COH associations.

The developments described in Chapter 8 dealing with the release of atomic nitrogen supplied a pattern for further inquiry. When mixtures of sucrose and copper sulfate were prepared and osmotized the observed increases in solution volume obtained at the end of the seven days required were precisely the sums of the increases effected by the involved solutes when osmotized individually. It was obvious, therefore that no interaction took place.

There followed a series of researches in which sucrose was paired with various salts and the mixtures were osmotized. In no mixture was there obtained any evidence of a disruption of the solute sucrose units obtained in the simple osmosis of sucrose alone.

A STUDY OF UREA

For what might be termed historically sentimental reasons urea as the organic compound first synthesized in the laboratory was chosen to represent a group of substances subject to characterization as non-electrolytes having carbon-nitrogen linkages. According to conventional chemistry the compositional formula for urea commonly was written as NH_2CONH_2, a formula which also had significance with respect to structure. The structural formula for urea commonly was written as follows:

$$\begin{array}{ccccc} H & & & & H \\ \diagdown & & & & \diagup \\ N & - & C & - & N \\ \diagup & & \| & & \diagdown \\ H & & O & & H \end{array}$$

The molecular weight of urea as above represented was reported as being 60.5, a weight based upon a value other than 2 for the weight of a neutral atom of hydrogen. As prescribed by the repeatedly-validated description of periodic hydrational potentiality the molecular weight of a substance with the indicated composition would be 64.

In conventional chemistry it has not been a common practice to project specific types of ionization for the component atoms of molecules. Such a projection, however, had been indicated as most important and helpful in numerous appraisals involved in preceding chapters, and especially so in the case of the carbon atom, as noted in the study of sucrose. A failure to project for the carbon atom a specific type of ionization was recognized as in practice equivalent to an assignment of covalent potentialities. It is of course quite possible that such potentialities exist, but up to this writing no evidence has been obtained in these researches to support such a viewpoint. Until such evidence appeared, therefore, covalency was not to be considered as a potential attribute of ionized atoms.

The study of sucrose had evidenced the incidence of carbon as positve ions and as negative ions. With respect to urea there thus was the suggestion that two types of ions were involved. These two projected types have been indicated in Diagram 4.

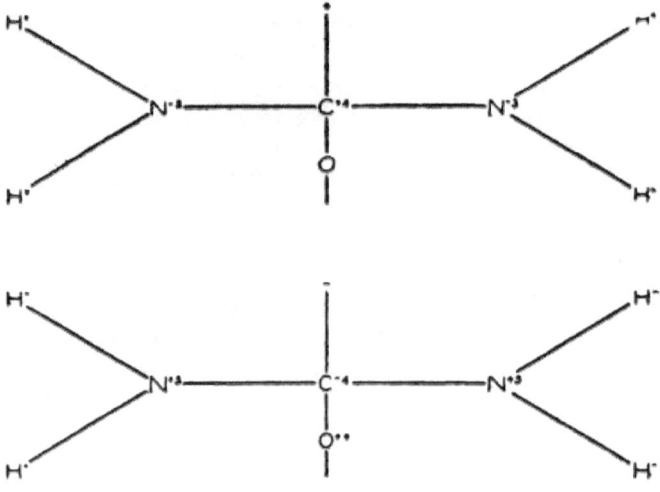

DIAGRAM 4 PROJECTED CONFIGURATION FOR IONIC COMPONENTS
OF UREA

It was obvious that these two ions readily could be projected as subject to union to form a balanced neutral molecule, that their amphoterism would account for their failure to conduct electricity and that their integrity as ions would render them subject to hydrational potentialities as prescribed by the description so extensively validated. It followed, therefore, that as ions their weights, equal as 64, would provide an ability to take on and hold 14 H_2O^- units in hydration. The data of Chapter 10 had indicated rather clearly that hydrational bondages could extend into the crystalline state. Most definitely it was recognized that projections involving errors would not be expected to lead to successful prediction or interpretation.

In the case of copper sulfate it was evidenced that three Cu^{++} ions shared 15 H_2O^- units and accounted for characterization as a "pentahydrate crystal." In the case of urea it was projected that the two mentioned amphoteric isomeric ions in crystallization shared 14 H_2O^- units. If this projection was soundly based there was every reason to anticipate not only a correlation with the observational data of specific gravity for aqueous solutions but also a correlation with osmotic behavior. The developments proved to be most interesting.

In several previous chapters the formula $d = 1.0 + \dfrac{Wa}{Wh}$ had been used in conjunction with specific gravity. With respect to urea it was to be noted that in the event that the crystalline molecule was composed of the two indicated ions and 14 H_2O^- units, representing the prescribed hydrational potentialities on a shared basis, the specific gravity was subject to calculation as $1.0 + \dfrac{2 \times 64}{2 \times 64 + 14\ H_2O^-}$, or $1.0 + \dfrac{128}{128 + 252}$, or 1.337. As reported in handbooks of tabulated data the specific gravity of crystalline urea was noted as 1.335. It was obvious that the specific gravity of the product was consistent with the projected composition.

It was to be assumed that in aqueous solution the ions indicated in Diagram 4 were released and became hydrated to the extent prescribed by the description of periodic hydrational potentiality. On this assumption one could calculate the expected specific gravity values for aqueous solutions at various concentrations and compare the values thus calculated with observed values. The data given in Table 45 provide for such a comparison.

The data given previously had evidenced the reliability of specific gravity as an index of the nature and status of solute units. In the

Table 45

Data affording a comparison of calculated and observed specific gravity values for aqueous solutions of urea.

Gms. Crystalline Urea per liter	Calculated ml hydrated Urea per liter ÷1.337	Calculated ml Solvent per liter Subtracted from 1000	Calculated sp. gr. of aqueous solution $\dfrac{\text{Col. 1 + Col.3}}{1000}$	Observed sp. gr. of aqueous solution 25°C Hydrometer
64	47.9	952.1	1.0161	1.016
128	95.7	904.3	1.0323	1.032
192	143.6	856.4	1.0484	1.050
256	191.4	808.6	1.0646	1.065
320	239.3	760.7	1.0807	1.080

Table 46

Data affording a comparison of the observed osmotic behavior of urea with the behavior calculated on each of two assumptions regarding the involved ions.

Assumed Ions	W_a	n	H	W_h	V_h	V_a	D_1	$\frac{D_1}{2}$	$\frac{V_a}{2}$	D_2	u	V_1	V_t	Observed Standard Osmotic Increase
$(NH_2)_2CO^{+-}$	64	2	14	316	263	32	231	115.5	16	99.5	1	99.5		
$(NH_2)_2CO^{-+}$	64	2	14	316	263	32	231	115.5	16	99.5	1	99.5	199	200
NH_2^+	20	1	13	254	236	10	226	113	5	108	2	216		
NH_2^-	16	1	15	286	271	8	263	131.5	4	127.5	2	255		
CO^{+-}	28	1	9	190	166	14	152	76	7	69	2	138	609	200

case of urea the data of Table 45 were interpreted as evidence of the validity of the two projections given in Diagram 4 and the correlated representation of the crystalline solid as of the composition $2NH_2$-$CONH_2$. $14H_2O$. As thus represented, solubility in water involved the separation of the two respective amphoteric isomeric ions which were held both by chemical bondage and by hydrational bondage through the sharing of $14H_2O^-$ units. Incident to this separation each of the involved ions was evidenced as hydrating to the extent prescribed by the description of periodic hydrational potentiality in the presence of adequate aqueous solvent.

Fortunately, the preceding researches on osmosis had supplied an adequate background for a supplementary appraisal of urea as characterized from the data of specific gravity. Referring to Diagram 4 one could project that in osmosis the two types of ions would remain intact, as had been evidenced in the osmotic behavior of the nitrate and ammonium radicals, or that in osmosis the carbon-nitrogen linkages would be broken, as had been evidenced for the carbon-carbon linkages in the osmotic behavior of sucrose. Calculations based on each of these projections were made, and the results thus obtained were compared with observational data. Provision for such a comparison has been made in Table 46.

The data given in Table 46 were interpreted as evidence that there was no osmotic breakage of the carbon-nitrogen linkage in urea. This development thus was to be considered consistent with the osmotic behavior of the nitrogen—containing NH^+_4 and NO^-_3 radicals, which also were evidenced as having remained intact during osmosis. The development was in contrast to the osmotic behavior of sucrose, which evidenced an osmotic disruption of the carbon-carbon linkages.

As noted in Chapter 8, atomic nitrogen was released from the ammonium and nitrate radicals through joint osmotization with copper sulfate. It became of interest, therefore, to determine whether or not copper sulfate would be able to disrupt the carbon-carbon-nitrogen linkages in urea. The data obtained in such a study have been assembled to comprise Table 47.

The data given in Table 47 were interpreted as having evidenced very clearly that in osmosis copper sulfate effected the disruption of the carbon-nitrogen linkages in urea. It was made evident also that, as in the case of sucrose, the disruption of the linkages was not followed by the dissociation of resultant radicals. There thus was the implication that COH and NH_2 groups could be utilized in protoplasmic synthetic processes, an implication which appeared consistent with much analytical work in organic chemistry.

Table 47

Data affording a comparison of the observed osmotic behavior of an indicated mixture (Column O) with calculated values (Column Vt).

Assumed Ions	Wa	n	H	Wh	Vh	Va	D_1	$\frac{D_1}{2}$	$\frac{Va}{2}$	D_2	u	V_1	Vt	O
Cu++	62	2	15	332	280	31	249	124.5	15.5	109	1	109		
S+6	44	1	1	62	36	22	14	7	11	-4	1	-4		
O--	12	1	17	318	306	6	300	150	3	147	4	588		
(NH₂)CO-+	64	2	14	316	263	32	231	115.5	16	99.5	2	199	792	1300
Cu+5	62	2	15	332	280	31	249	124.5	15.5	109	1	109		
S+6	44	1	1	62	36	22	14	7	11	-4	1	-4		
O---	12	1	17	318	306	6	300	150	3	147	4	588		
NH+₂	20	1	13	254	236	10	226	113	5	108	2	216		
NH-₂	16	1	15	286	271	8	263	131.5	4	127.5	2	255		
CO+-	28	1	9	190	166	14	152	76	7	69	2	138	1302	1300

CHAPTER 18

HYDRATION AND ATOMIC WEIGHT

More than half a century ago such researches as those of Pfeffer on osmosis and of Kohlrausch on electrical conductivity, allied with the interpretation of Vries and Hoff, brought forward a recognition of the analogy between solutes and gases. The analogy stemmed primarily from the relative isolation and freedom of movement of the involved atomic or molecular units within their respective environments. The gas laws had attained eminence among the mathematical descriptions of science and the nicety with which they characterized the behavior patterns of gases was attributed to the collective activity of large numbers of individual units. For a time there was a prospect that the collective activity of large numbers of solute units would mediate the development of similarly satisfying solution laws. The Kohlrausch law of the independent migration of ions afforded optimism, but soon it was concluded that the description was applicable only at low solute concentrations, since the apparent reduction in specific molecular electrical conductivity became interpreted as evidence of incomplete dissociation. During the subsequent years the theory of incomplete dissociation attained widespread popularity, with an accompanying natural deprecation of the evidence for the persistence of solute units with increase in solute concentration. The description of periodic hydrational potentiality as given and repeatedly validated in the considerations of preceding chapters constituted a basis for a re-evaluation of solution data.

Insofar as the electrical conductivity of aqueous solutions was concerned it became obvious that hydration brought about a transfer of H_2O^- units from solvent to solute and thereby rendered a prepared solution more concentrated than had been recognized. The involved error in appraisal was continued but progressively reduced in successive dilution and this development was subject to correlation with the apparent increase in specific molecular electrical conductivity with decrease in solute concentration. The data of specific gravity as given in Chapter 3 and 9 clearly evidenced for the involved solutions a general persistence of independence for the contained solute units. Hydrational bondage obviously represented a restraining force which under electrical stress in the absence of free solvent readily might be projected as interfering with movement and electrical conductivity, but in such cases there was involved a departure from the area of solutions. Thus there was cumulative evidence that in aqueous solutions there was no incomplete dissociation with increase in solute concentration. This was held to be a very important consideration, especially in relation to osmotic behavior and the recognition of solute behavior patterns interpreted as attributable to the collective activity of large numbers of relatively free and independent units with a consequent analogy with gases.

In preceding chapters the behavior patterns of isolated solute ions in relation to specific gravity and osmosis uniformly have evidenced the validity of the weight values prescribed by the description of periodic hydrational potentiality and its corollaries. Perhaps ordinarily these behavior patterns might constitute widely acceptable documentary evidence of the integrity of the numerous involved atomic weight values, but in the case of chemical science there existed a formidable background of traditional viewpoints instinctively oriented to support the present assemblage of irregular fractional atomic weight values for the chemical elements. It was recognized that the conventional atomic weight values represented interpretations of data obtained laboriously over a period of many years with meticulous attention to detail. It was recognized also that in recent years supplementary data obtained with the mass spectrograph had been interpreted as signifying the presence of neutrons in atoms in numbers appropriate to account for and therefore sustain a validity for the contemporary atomic weight values of the respective kinds of elements. For example, the element uranium, heaviest of the naturally-occurring elements, had been interpreted as possessing a nucleus containing 92 protons and 146 neutrons, values introduced as accounting for a summation atomic weight of 238 for the element. It was interesting and significant that the various behavior patterns of the solute uranyl radical and the solute uranium ion failed to sustain any such value as 238 for the atomic weight of uranium. There was no evidence from the behavior patterns of solute element ions to sustain any irregular fractional atomic weight value assigned by chemical science. On the contrary, the behavior patterns evidenced the integrity of the values prescribed by the description of periodic hydrational potentiality. Under these circumstances it appeared necessary and desirable to re-examine observational data historically basic to the derivation of the irregular fractional values long accepted as atomic weights.

In substantial measure the exhaustive and exhausting researches on the determination of atomic weight values for the chemical elements were reviewed, summarized and appraised in various publications and editions by Frank Wigglesworth Clarke. For the purposes of this study reference was made to the work entitled "A Recalculation of the Atomic Weights" (Fourth Edition, Revised and Enlarged) which occurred in Memoirs of the National Academy of Sciences, Volume 16, pp. 5-418, 1922. In this publication the data which appeared to be the most reliable were selected, cited and analyzed statistically with a view to the procurement of one atomic weight value for each element, this one value to represent maximal accuracy.

Anyone confronted with this publication could not fail to be impressed by the amount of energy which had been directed to the obtaining of the original data and to the involved treatment of the data. Nevertheless it was to be recognized that from the very nature of the procedures no thought was given to the possibility of change in weight with ionization,—to the possible existence of categories of weight as represented in the corollaries of the description of periodic hydrational potentiality and the consequent possibility that different ionic forms of the same element might possess different and characteristic weights. The procedures therefore reflected the ancient concept of atoms as solid balls and involved no consideration of the subsequent recognition of their electrical nature. In substantial measure also the data involved end products often in a solid state and always subject to appraisal as having been weighed in a neutral state. Except in relation to helium and its chemical analogs the procedures and data were not adapted to the appraisal of elements as isolated neutral or ionic atoms.

The procedures designed to yield atomic weight values often involved chemical analyses. In such cases wherever feasible the principal reference value in use was $O = 16$, a value which, as mentioned previously, almost certainly had been derived by inference from the value $O_2 = 32$ at a time when the solid ball concept of atoms prevailed. Whenever oxides of specific elements were available it was customary to use these compounds in the analyses and to a great extent the conventional atomic weight values ascribed to the heavier elements directly or indirectly reflected the use of the value $O = 16$ as representing the weight of the oxygen atom when in combination as the bivalent anion O^{--}. Data given in numerous preceding tables evidenced a weight of 12 for such an ion. It was recognized that with 16 instead of 12 as a basic weight value for oxygen the relative atomic weight values derived by ratios for associated elements would be higher, particularly for the heavier elements. Because of this fact attention was directed to atomic weight values which had been derived for some heavier elements through the use of gravimetric data involving the composition of oxides. Some results obtained have been given in Table 48.

In the data of Table 48 the conventional atomic weight values cited were obtained typically from the indicated ratios,—ratios which represented averages of values obtained in various determinations by various investigators. It was noted that ratios calculated through the use of the prescribed weight values for ions specified in the column on the right were of the same order of magnitude. This meant that the gravimetric data so painstakingly derived in efforts to obtain one and only one atomic weight value for each element irrespective of its state lent themselves nicely to an alternative interpretation in validation of the values prescribed by the description of periodic hydrational potentiality and its corollaries. With respect to oxygen it meant that the bivalent anion O^{--} as a free ion could be interpreted as having had a weight of 12, as had been repeatedly evidenced by solute behavior, and that both this weight and the combining weight of 14 had become obscured through the involvement of the neutral end product O_2. In keeping with the data of foregoing chapters this appeared to constitute the preferred most likely interpretation.

The procedures designed to yield atomic weight values for the elements often involved compounds and chemical reactions in which oxygen was not present. Under these circumstances other elements served as secondary standards of weight in the place of oxygen and among such elements silver was accorded prominence. The relative combining weight of silver atoms as represented in the oxide Ag_2O was subject to calculation from the description of periodic hydrational potentiality and its corollaries as 95 and the combining weight of the involved oxygen anions would be 14. The molecular weight of the silver oxide Ag_2O would be calculated as 204. The ratio of the weights of the two elements in the compound on an atomic basis would be calculable as

$$14 : \frac{204 - 14}{2} , \text{ or } 14 : 94$$

One may venture to surmise that conventionally the ratio was projected for gross weights in the manner $14 : 94 :: 16 : \times$ to yield for the derived atomic weight of silver such a value as 108.57. As thus projected the use of silver as a secondary standard of weight may be interpreted as having contributed further to the origin of irregular fractional atomic weight values in excess of the regular integral values prescribed and evidenced to such an extent in preceding chapters.

Table 48

Data affording a comparison of observed ratios involved in the derivation of conventional atomic weight values with ratios calculated from the description of periodic hydrational potentiality and its corollaries.

Element	Conventional Atomic Weight	Prescribed Weight of Neutral Atoms	Involved Conventional Gravimetric Data	Averages of Conventionally Found Ratios	Ratios Derived from Prescribed Wts.	Prescribed Weights for Elements Involved in the Conventional Gravimetric Data
Lanthanum	138.92	114	$\dfrac{La_2O_3}{La_2(SO)_3}$	.534	.535	$La^{+3}=120$ $O^{--}=12$ $S^{+6}=44$
Tungsten	184.0	148	$\dfrac{W}{WO_3}$	.7931	.816	$W^{+6}=160$ $O^{--}=12$
Mercury	200.61	160	$\dfrac{Hg}{NgO}$	.92595	.932*	$Hg^{++}=164$ $O^{--}=12$
Lead	207.21	164	$\dfrac{Pb}{PbO}$	.928271	.933*	$Pb^{++}=168$ $O^{--}=12$
Bismuth	209.00	166	$\dfrac{Bi}{Bi_2O_3}$	.896683	.896	$Bi^{+3}=172$ $O^{--}=12$
Uranium	238.07	184	$\dfrac{UO_2}{UO_2C_2O_4}$	.71782	.7105	$U^{+6}=196$ $O^{--}=12$ $C^{+4}=20$

*See also Table 11

Data pertaining to some ratios involving silver in relation to the conventional derivation of irregular fractional atomic weight values have been given in Table 49.

With respect to the gravimetric data for the composition of oxides as given in Table 48 the procedures commonly involved the release of ionic oxygen and the projected ratios reflected this state. In contrast, with respect to the gravimetric data represented in Table 49 the chemical procedures involved reactions of substitution or exchange and in consequence the ratios reflected weight values characterizing the neutral state. It was to be noted that the observed gravimetric ratios again lent themselves nicely to an alternative interpretation to yield atomic weight values prescribed by the description of periodic hydrational potentiality and its corollaries. Collectively the elements included in the two tables were appraised as fairly representative of the system of chemical elements and in view of the behavior patterns considered in preceding chapters there was no doubt but that the alternative interpretation was the one to be favored.

It was recognized that the niceties of correlation which were made evident in the tabulated data for specific gravity and osmosis documented for solute units a homogeneity definitely inconsistent with the variations which characterized the gravimetric data and allied ratios as given in the work of Clarke. The ratios cited by Clarke as averages stemmed from values whose range of variation thereupon became of interest in conjunction with atomic weight. It was obvious that the description of periodic hydrational potentiality provided a basis for prescribing definite limits of variation consistent with the specific types of units subject to easy and natural projection as having been involved. On this account it became of interest to compare some actual found gravimetric ratios with ratios representing the limits prescribed for the specified units. Data affording such a comparison have been given in Table 50, A and B.

The data of Tables 50A and 50B relate to the relative weights of elements in the neutral and cationic states in the indicated respective oxides. The arrangement of the data has been such as to permit the change in weight prescribed by the description of periodic hydrational potentiality and its corollaries to afford an explanation of the order of diversity represented by the gravimetric data of chemical science. To an appreciable extent the conventionally derived variations in ratios therefore could be interpreted as documenting the prescribed change in weight with ionization. It would seem sufficient, however, to point out that in every listed oxide the found weight ratios fell within the limits subject to calculation as represent-

Table 49

Data affording additional comparisons of observed ratios involved in the derivation of conventional atomic weight values with ratios calculated from the description of periodic hydrational potentiality and its corollaries.

Element	Conventional Atomic Weight	Prescribed Weight of Neutral Atoms	Involved Conventional Gravimetric Data	Averages of Conventionally Found Ratios	Ratios Derived from Prescribed Wts.	Prescribed Weights for Elements Involved in the Conventional Gravimetric Data
Barium	137.36	112	$\dfrac{BaBr_2}{2AgBr}$	1.37745	1.34	$Ba° = 112$ $Br° = 70$ $Ag°_D = 94$
Cerium	140.13	116	$\dfrac{CeCl_3}{3Ag}$	0.76167	0.773	$Ce° = 116$ $Cl° = 34$ $Ag° = 94$
Tantalum	180.88	146	$\dfrac{TaCl_5}{5AgCl}$	0.49611	0.494	$Ta° = 145$ $Cl° = 34$ $Ag° = 94$
Osmium	191.50	152	$\dfrac{K_2O_2Cl_6}{4AgCl}$	0.84097	0.844	$O_5° = 152$ $K° = 38$ $Cl° = 34$ $Ag° = 94$
Platinum	195.23	156	$\dfrac{K_2P_TCl_6}{4AgCl}$	0.84809	0.85	$P_T° = 156$ $K_5° = 38$ $Cl° = 34$ $Ag°' = 94$
Gold	197.2	158	$\dfrac{Au}{AgBr}$	0.95222	0.964	$Au° = 158$ $Br° = 70$ $Ag° = 94$

166

Table 50A

Data affording a comparison of calculated and observed weight ratios.

Element and Atomic Number	Involved Compositional Ratio	Prescribed Weight Ratio For Neutral Atoms Wn	For Ionic Atoms Wa	Observed Weight Ratios Conventionally used for the Determination of Atomic Weight Values and Reference	No. of Tests	Conventional Derivation	Atomic Weight Values Prescribed Neutral Atoms	Prescribed Ionic Atoms
Iron 26	$\dfrac{Fe_2}{Fe_2O_3}$	.684	.724	.6931 to .7007 Clarke, pp. 321–323	55	55.883	$Fe°=52$ $O°=16$	$Fe^{+3}=58$ $O^{--}=12$
Cobalt 27	$\dfrac{Co}{CoO}$	.771	.830	.78508 to .78859 Clarke, pp. 330–341	57	58.956	$Co°=54$ $O°=16$	$Co^{++}=58$ $O^{--}=12$
Nickel 28	$\dfrac{Ni}{NiO}$	.778	.834	.78232 to .79135 Clarke, pp. 329–343	80	58.676	$Ni°=55$ $O°=16$	$Ni^{++}=60$ $O^{--}=12$
Copper 29	$\dfrac{Cu}{CuO}$	.784	.838	.79770 to .80099 Clarke, pp. 125–127	24	63.561	$Cu°=58$ $O°=16$	$Cu^{++}=62$ $O^{--}=12$
Zinc 30	$\dfrac{Zn}{ZnO}$	.790	.842	.80274 to .80570 Clarke, p. 179	6	65.397	$Zn°=60$ $O°=16$	$Zn^{++}=64$ $O^{--}=12$
Selenium 34	$\dfrac{Se}{SeO_3}$	.680	.760	.7100 to .71250 Clarke, pp. 292–296	28	79.168	$Se°=68$ $O°=16$	$Se^{+4}=76$ $O^{--}=12$

Table 50B
Data affording a comparison of calculated and observed weight ratios.

Element Involved and Atomic Number	Compositional Ratio	Prescribed Weight Ratio		Observed Weight Ratios Conventionally used for the Determination of Atomic Weight Values and Reference	No. of Tests	Conventional Derivation	Atomic Weight Values Prescribed Neutral Atoms	Ionic Atoms
		For Neutral Atoms W_n	For Ionic Atoms W_a					
Molybdenum 42	$\frac{Mo}{MoO_3}$	.636	.728	.66495 to .66741 Clarke, pp. 273-275	28	96.03	$Mo^\circ=84$ $O^\circ=16$	$Mo^{+6}=96$ $O^{--}=12$
Ruthenium 44	$\frac{Ru}{RuO_2}$	.734	.800	.76023 to .76075 Clarke, pp. 345-346	10	101.65	$Ru^\circ=88$ $O^\circ=16$	$Ru^{++}=96$ $O^{--}=12$
Cadmium 48	$\frac{Cd}{CdO}$	.858	.894	.87491 to .87518 Clarke, p. 190	9	112.38	$Cd^\circ=96$ $O^\circ=16$	$Cd^{+2}=100$ $O^{--}=12$
Tellurium 52	$\frac{Te}{TeO_2}$	.764	.823	.79625 to .80207 Clarke, pp. 297-310	76	127.53	$Te^\circ=104$ $O^\circ=16$	$Te^{+4}=112$ $O^{--}=12$
Tantalum 73	$\frac{Ta}{Ta_2O_5}$	.784	.838	.81868 to .81959 Clarke, p. 262	5	181.3	$Ta^\circ=146$ $O^\circ=16$	$Ta^{+5}=156$ $O^{--}=12$
Bismuth 83	$\frac{Bi_2}{Bi_2O_3}$	.874	.906	.89647 to .897035 Clarke, pp. 256-259	49	208.06	$Bi^\circ=166$ $O^\circ=16$	$Bi^{+3}=172$ $O^{--}=12$

ing transition from the neutral to the indicated ionic state. In general the heavier the cationic element the more commonly was a tendency to remain ionic suggested or made evident. Alternatively expressed, with increase in weight there was a suggested or evidenced reduction in the ability of the elements to obtain or retain orbital electrons. The development seemed of interest in relation to the radio-active elements. The important consideration at this point, however, was the fact that in all of the more than four hundred analyses involved the observed gravimetric ratios fell within the limits prescrib ed by the precise change in weight with ionization which repeatedly had been validated in the data of preceding chapters.

It was quite obvious from the data of the three preceding tables that the observational gravimetric data obtained by chemists in procedures renowned for their attention to detail and the ratios derived from these data were subject to interpretation in a manner to document the integrity of the first three corollaries of the description of periodic hydrational potentiality. Yet with this knowledge it had become quite as obvious during the past thirty years that the mere suggestion of inaccuracy with respect to the conventional irregular fractional atomic weight values was exceedingly provocative. At this point, therefore, attention will be directed to some facts outside the area of precise measurement.

There is worldwide recognition among educated people that chemical science has played a leading role in the development of civilization. Its contributions have been beyond enumeration and appraisal,— and they reasonably may be expected to continue It has been natural and justifiable that under these circumstances chemistry has become a proud science and profession. The indoctrination involved in the training of chemists has imbued the atomic weight values of chemistry with an axiomatic assumption of integrity tending to decry beyond all reason any suggestion of error. Beyond that one scarcely would expect unalloyed pleasure to come with the discovery that long-treasured jewels were but paste.

Yet if pleasure is a reasonable goal one may doubly rejoice : once for the alternative interpretation of the chemist's gravimetric data yielding a mathematical basic order, and once to receive chemists into the glorious order of human beings subject to error. The adjustments may not prove particularly difficult. The data obtained from mass spectrographic analyses certainly evidence the presence of units of different integral weights, and these differences may be appraised as reflecting the unusual applied stresses and the consequent production of heterogeneous types not stable under the conditions of solution.

Such an appraisal would explain, for example, the heterogeneity conventionally accorded the element chlorine. Aqueous solutions of sodium chloride attest the homogeneity of the solute chlorine ion Cl^-, and the composition of such radicals as ClO_3^- and ClO_4^- attests the pre-existence of the ions Cl^{+5} and Cl^{+7} respectively. It follows that under appropriate stresses chlorine would yield ions of appreciable heterogeneity.

With respect to solutes it became obvious that in the absence of an adequate description of hydrational potentiality the behavior patterns of solute ions of necessity would remain obscure. It was just as obvious, however, that with the attainment of such a description the behavior patterns of solute units supplied with a nicety comparable with that characterizing the gas laws not only the documentation of basic order but also the background for adequate descriptions as solution laws. Emergent from such a background is a recognition of hydrated ions as representing a short of intermediate state between gases and liquids, and yet within the scope of preceding chapters hydrational potentiality has been evidenced as present among gases and solids as well as in solution. Hydrational potentiality therefore may be appraised as an almost universal phenomenon in nature.

From the standpoint of utility the outstanding feature of the periodicity of chemical attributes and behavior patterns after Mendeleev was the facility with which the description mediated successful prediction. From the standpoint of science facility for successful prediction constitutes for any description the most critical index of integrity, since it signifies an inherent mathematical relationship or function. Repeatedly in the data of this and preceding chapters the description of periodic hydrational potentiality has exemplified facility for successful prediction.

In the free world there still persists within limits a certain freedom of thought and one may choose to interpret data in a manner compatible with ability to appraise its significance. For some the long revered conventional irregular fractional atomic weight values will have become endowed with a sanctity approaching that commonly accorded to imagined supernatural ghosts. No one would presume to deny these the right to cling to the values. For others, however, there will come a recognition of regular integral atomic weight values and a consequent recognition and appreciation of basic order at the atomic and molecular levels so widespread and important in inorganic reactivity and the superimposed elaborations characterizing the organic world.

It has seemed appropriate to report at this point on some results obtained in a study of uranium. After many years as an element of academic interest in the annals of chemical science the researches of the Curies initiated for uranium an era of dynamic importance. Largely because of the Manhattan project involving it in the release of nuclear energy uranium attained widespread popular as well as scientific interest.

In conjunction with the notable developments relating to uranium considerable emphasis has been placed on atomic weight values. This has been especially true in the case of a fissionable state, whose designation as U 235 represented atomic weight as a specific characterization.

The description of periodic hydrational potentiality given and extensively validated in preceding chapters prescribed the value 184 as the weight of uranium atoms in the neutral state, the value 196 as the weight of the 6-valent positive ions of uranium and the value 190 as the combining weight of the U^{+6} ion. The description also prescribed an absence of hydrational potentialities for neutral and cationic uranium. It became of interest to study some behavior patterns of uranium with respect to compliance with the description of periodic hydrational potentiality.

For reasons previously indicated specific gravity was considered to be the most reliable index of the nature and status of the solute ions present within an aqueous solution. The attribute particularly was valuable whenever considerations of ionic weight were involved because commonly the tabulated observational data of specific gravity for aqueous solutions as obtained by various investigators over a period of years reported solute concentrations based upon grams per liter without regard to the molecular weight of the solute.

It was held to be possible but highly improbable that from erroneous assumptions in regard to the nature and status of solutes successful calculations or predictions could be made. On this premise uranyl nitrate was selected as a suitable uranium compound for special study.

It was found that no satisfactory order of agreement between calculated and observed specific gravity values was obtained if the assumption was made that the ions present per introduced solute molecule of uranyl nitrate were UO_2^{++}, NO_3^- and NO_3^-. This negative result in itself was interpreted as impressive evidence of the value of specific gravity as an index, since it carried the suggestion that if the uranyl radical UO_2^{++} had been present in the compound it had become dissociated incident to the dissolving of the compound in water.

It was found that a satisfactory order of agreement between calculated and observed specific gravity values was obtained if the assumption was made that the ions present per unit molecule of uranyl nitrate were U^{+6}, O^{--}, O^{--}, NO_3^- and NO_3^-. Data affording a comparison of calculated and observed specific gravity values under such conditions were assembled to comprise Table 51.

In Table 51 the values for the initial grams of solute per liter and for the observed specific gravity of the aqueous solution were taken from the International Critical Tables and there appeared to be no reason for questioning their validity. The minus values for the calculated milliliters of solvent present represented evidence and measure of the extent of hydrational bondage, a subject discussed at some length in preceding chapters. The indicated order of agreement between calculated and observed values was interpreted as beyond chance. It was recognized that any projected departure from the assumptions made with regard to the solute ions present or from the W_a values for their weights in the anhydrous state would involve departures from the satisfactory order of agreement. The tabulated data therefore were appraised as having supplied impressive evidence documenting the integrity of the value 196 as the weight of the uranium U^{+6} ion.

The data of Table 51 were considered to be of further interest in relation to the historical derivation of the atomic weight value in present use in chemical science. If 16 was substituted for 12 in the W_a column and the weight ratio of oxygen to uranium was established as $12:196::16:X$, the value of X became 261. If the combining weight of the O^{--} ion was substituated for the ionic weight and the ratio was established as $14:196::16:X$, the value of X became 217. Theoretically, in compliance with the description of periodic hydrational potentiality the values 217 and 261 represented a range of expected variation contingent upon the use of the value 16 representing the weight of the oxygen atom in the neutral state. The mean value, 239, was of the order of the conventional atomic weight value for uranium, a value which also represented an average. These relationships were considered very suggestive with respect to the derivation of the conventional atomic weight value for uranium.

In several preceding chapters the osmotic behavior of solutes was demonstrated as being of value as an index of the nature and status of the involved solute ions. It became of interest, therefore, to study the osmotic behavior of uranyl nitrate. It had been noted repeatedly in previous studies of the osmotic behavior of nitrates that

Table 51

Data affording a comparison of calculated and observed values for the specific gravity of aqueous solutions of uranyl nitrate.

Assumed Ions	Wa	Wh	$\frac{Wa}{Wh}$	Average $\frac{Wa}{Wh}$	Original Grams Solute per Liter	Calculated Grams Hydrated Solute	Calculated Ml Hydrated Solute	Calculated Ml Solvent	Calculated Specific Gravity of Solution	Observed Specific Gravity of Solution
U+6	196	Anhydrous	1.000		586.4	2063	1610	−610	1.458	1.466
O−−	12	318	0.0377		629.16	2220	1732	−732	1.492	1.498
O−−	12	318	0.0377	0.2839	673.64	2370	1847	−847	1.523	1.532
NO₃⁻	60	348	0.1732		720.82	2540	1975	−975	1.565	1.567
NO₃⁻	60	348	0.1732							

Table 52

Data affording a comparison of calculated and observed osmotic behavior patterns for uranyl nitrate.

Assumed Ions	Wa	n	H	Wh	Vh	Va	Di	$\frac{Di}{2}$	$\frac{Va}{2}$	Vi	Calculated Standard Osmotic Increase	Observed Standard Osmotic Increase
U+6	196		0			98			49	49		
O−	12	1	17	318	306	6	300	150	3	147		
O−	12	1	17	318	306	6	300	150	3	147		
NO₃⁻	60	2	16	348	297	30	267	133.5	15	118.5		
NO₃⁻	60	2	16	348	297	30	267	133.5	15	118.5	580	580

the NO_3^- radical remained intact except under special conditions. It was assumed, therefore, that the osmotic behavior of uranyl nitrate would involve the solute ions itemized in Table 51. On this basis the calculated osmotic behavior and the observed osmotic behavior yielded the results brought together for comparison in Table 52.

The data given in Table 52 were interpreted as having verified with a very satisfying degree of nicety the integrity of the assumptions regarding the solute ions present in aqueous solutions of uranyl nitrate and of their respective projected patterns of behavior in osmosis. The osmotic behavior patterns calculated on the basis of alternative assumptions were noted as appreciably divergent. For example, the calculated value for UO_2^{++}, NO_3^- and NO_3^- was 292, and that for the component atoms as element ions was 1401. Osmotic behavior, therefore, again was evidenced as being a very valuable tool in research.

Collectively the data of Tables 51 and 52 were interpreted as having documented the validity of the value 196 as the weight of the uranium U^{+6} ion,—a value in complete accord with the repeatedly evidenced integrity of the description of periodic hydrational potentiality.

REFERENCES

1. CLARKE, FRANK WIGGLESWORTH (1922). A Recalculation of the Atomic Weights. (Fourth Edition, Revised and Enlarged) Memoirs of the National Academy of Sciences, Volume 16. pp. 5—418.
2. HOFF, J. H. VAN'T (1887). Die Rolle der osmotischen Druckes in der Analogie Zwischen Losungen und Gasen. Zeitschr. Physiol. Chem. 1:481-508.
3. KOHLRAUSCH, F., & H. VON STEINWEHR (1902). Weitere untersuchungen uber das Leitvermogen von Elektrolyten aus einwerthigen Ionen in Wasseriger Losung. Akad. d. Wiss. Berlin Sitzungsberichte 581-587.
4. VRIES, H. DE (1884). Eine Methode zur Analyze der Turgorkraft. Jahrt. Wiss. Bot. 14:427-601.

CHAPTER 19

HYDRATION AND THE INORGANIC WORLD

In keeping with time-honored precepts in science the context of preceding chapters has maintained a more or less substantial contact with observational data. Granted that this standard practice has proven eminently desirable to discourage the complete substitution of dreams for facts it is nevertheless true that most of what presently is designated as knowledge was once housed within the category of dreams. In these closing chapters of this book, Chapters 19 and 20, there will be no particular allegiance to observational data, on which account, in the absence of any imposed mental stress for appraisal, it is hoped that there may prevail a mood for carefree adventure.

The description of periodic hydrational potentiality by virtue of ts intimate integration with the ninety-two naturally-occurring elements appears to have significance not only in relation to a revision of the atomic weight values and a substantiation of the integrity of the dream of Prout, but also in relation to the interesting problems involved in the origin of the elements and the origin of this and other solar systems. The description itself rather clearly carries the implication that moisture was involved in the evolution of the elements. Data given in Chapters 12 and 13 suggested or evidenced that gaseous ions possessed the identical hydrational potentialities which had been prescribed and evidenced for solute ions in the data of earlier chapters. Data given in Chapter 10 indicated that the same hydrational potentialities extended into the solid state. The presence of water during early geological periods has been evidenced in various ways. The presence of water vapor in still earlier geological periods commonly has been projected.

A common viewpoint regarding the origin of the earth is one which pictures it as having evolved from a maelstrom of hot gases, pre-eminent among which was the element hydrogen. This viewpoint naturally is of special interest from the standpoint of the Prout hypothesis. Data given in Chapter 8 suggested or evidenced the release of H^- ions of weight zero. Under the Graham Law of Diffusion an atom of weight zero would be assigned infinite mobility and would thus become subject to characterization as radiant energy. Such ions, considered as projected from the sun in all directions, would be subject to frictional electron loss upon impact. One consequence of such a loss would be the imposition of a negative charge, and in the case of the earth would afford an explanation of the creation and maintenance of the earth's negative charge. Another consequence of such a loss would be the acquisition of weight by the hydrogen atom, and in the case of the earth this development would implement mechanisms for the maintenance of the earth's gaseous ionospheres and for an accretionary evolution of the earth's mass. It is of interest also that the unusual properties of H^- ions, appraised as the most dynamic of the elemental states of matter, could be correlated speculatively with the geological concept of a maelstrom of hot gases through the forces of aggregation involved in interdiffusion to surmount the proton-to-proton barrier and thus mediate the build-up of the more complex elements. Commonly the temperatures projected for such syntheses have been exceedingly high, but these temperatures have been based upon forces outside the area of interdiffusion in which a mutuality of attraction is involved rather than a randomized impact.

The H^- ions have interest further in relation to astronomy. According to the famous formula of Einstein the relation of energy and mass is given as $e = mc^2$ when e represents energy, m represents mass and c represents the speed of light. This formula applied to H^- ions of zero weight would prescribe the radiation phase of their dispersal, as from a radiating source, as without energy, but as representing energy upon a contact sufficient to dislodge the electron and give to the resulting neutral hydrogen atom a weight of 2. The projection seems particularly pertinent in cosmic relations. In the presence of moisture the characteristic maximal hydrational potentiality of the H^- ions introduces the concept of H_2O^- aggregation and a synthesis of water. From a gravimetric standpoint an ion without weight which has taken on 23 H_2O^- units, each of weight 18, might be projected as representing water. Alternatively an approach to the atomic structure of water might involve an analysis of the force with which H_2O^- units are held by water. For the present the comparison of hydrational and electron bondages presents adequate challenge without a consideration of additional sub-atomic entities.

The Einstein formula is not without interest in relation to the speculative contemplation of the origin of the elements from hydrogen as projected by the Prout hypothesis. The Graham law of diffusion described it as a repulsive force acting to separate homogeneous units and in application the relative diffusion rate for a unit of zero weight would be calculable as infinity. The C^2 component of the Einstein formula could be projected as a substitute for infinity, in which case e would have a value approximating 3.77×10^7. On the same basis the force of interdiffusion would be subject to calculation as approximately one-half that value, or 1.885×10^7. This force would be approximately 18×10^7 times as great as the interdiffusion calculable for H° and H°_2. It follows that theoretically the force of interdiffusion operative between H^- ions and H° atoms would far transcend any force present within the ordinary realm of atoms, on which account it might be projected as having potentialities for the synthesis of heavier elements from hydrogen.

The successful tapping of nuclear energy permits the anticipation of experimental and eventual practical procedures for controlling the synthesis of heavier elements and the mathematical description of the involved behavior patterns. The future therefore seems to assure ample opportunities for the appraisal of the potential role of interdiffusion in the synthesis of elements.

For the present the synthesis of elements provides a delightful era of speculative probing. Out in the remoteness of the cosmos one may witness the birth of stars and may note further that the involved elements are among those comprising the earth. By comparisons one may conclude that our sun and its planets represent the attainment of an appreciable age with attendant changes. The considerations prompt curiosity as to whether the natural synthesis of the elements of our earth took place exclusively in a remotely early era, or whether synthesis is a continuing and contemporary process.

Perhaps rather surprisingly the observational data of a preceding chapter well may have an important bearing on the synthesis of elements. The data given in Chapter 8 appeared to document the integrity of H^- ions of zero weight. Mendeleev in Russia and Harkins in the United States had given prominence to the plausibility of zero weight: but they had associated zero weight with an unknown chemical element, and they had been completely frustrated by the seemingly impossible gravimetric approach of conventional chemistry. The description of periodic hydrational potentiality supplied an unexpected solution to the problem since it prescribed a potentiality for hydration and following hydration both the weight and the volume of the H^- ions were subject to measurement. Such measurements established by retroaction the integrity of H^- ions of no weight. These zero H^- ions in an anhydrous state when considered in conjunction with the laws of diffusion and interdiffusion as given in preceding chapters supplied a basis for a suggestion that the synthesis of elements is a continuing and still contemporary process.

With H⁻ ions in the anhydrous state appraised as a component of solar radiation there arises immediately the matter of their fate following interception by the earth. With the loss of the outer orbital electron, subject to speculative correlation with ionospheres and the earth's negative charge, there would result the formation of neutral atoms of hydrogen of weight 2. If the observational data of preceding chapters had given any support to the contemporary chemical concept that neutrons were adjuncts to atoms appraised as protons with orbital electrons an atomic infiltration by neutral hydrogen atoms might be projected. Since no such support was given, however, and the neutral atoms possessed weight a gravitational infiltration readily could be envisioned as moving these H° units into an inner-earth region characterized by conditions effecting the release of H⁻ ions, whereupon interdiffusion would bring about a synthesis of heavier elements. Such a projection would characterize synthesis as a continuing and still-contemporary process which in a general way would account for a progressive increase in weight and volume of the earth with an attendant reduction in its velocity and temperature. In addition to the obvious relation to the earth's crust the projection also would be consistent with the development of such internal stresses as have been documented by earthquakes, volcanoes, rifts, thrusts and dikes. The projection also would be consistent with an internal synthesis of radioactive elements appraised as representing instability imposed in a manner somewhat analogous to the modern production of radioactive isotopes.

The speculative aspects relating to synthesis on our earth appear as interesting in relation to a cosmic pattern. The positions of our sun's planets with respect to the sun document a mathematical sequence subject to correlation with interdependent forces, including synchronized adjustments to differential accretion. The rich endowment of astronomy has included contributions from researches in such areas as mathematics, optics, spectroscopy, physics and chemistry: in fact no area of science is self-sufficient. It would be interesting indeed if an extrapolation from the behavior patterns of solute ions, however fanciful and tenuous at present, should point out an improved pathway to the stars.

On a contrasting basis of meticulous measurement it has been the high privilege and responsibility of the data given in preceding chapters to document the attributes and behavior patterns of atoms as strictly mathematical. In some slight measure an analogous characterization of molecules has been documented; but the world of molecules is far too extensive to be seriously appraised on the basis of a very restricted sampling.

The tabulated data relating to the behavior patterns of atoms have included a representative number of the ninety-two naturally-occurring elements. The behavior patterns have been based upon atoms in free and combined states, in anhydrous and hydrated states, in gaseous, liquid and solid states, in neutral and ionic states, in positive and negative ionic states. The attributes involved have been weight, volume, density, mobility, electrical charge, neutrality, acid-alkali potentiality, chemical potentiality and hydrational potentiality. The processes involved have been diffusion, osmosis, interdiffusion, chemical bondaging, hydrational bondaging, crystal formation, dissociation and synthesis. From these data the conclusion has been drawn that without the slightest resort to indulgence in speculation the inorganic world is composed of atoms subject to characterization in integral mathematical terms documenting the revelation of a basic order hitherto appreciably obscurred by natural but unfortunate misinterpretations.

From such a background of basic order among the elements one may venture to integrate the order with the relatively recent developments involving nuclear energy. Arranged in the order of increasing atomic number the naturally-occurring radioactive elements evidence a progressive decline in the ability to capture and retain electrons. In the element uranium, the terminal element, this decline might be appraised as having reached a point at which the ability to capture and retain even a single electron had been greatly impaired. In uranium also there had been lost the ability to attract H_2O^- units in hydration. The principle of change in weight with ionization,—a principle contained within the repeatedly validated description of periodic hydrational potentiality—insured that the loss of electrons by neutral uranium atoms would result in an increase in weight beyond the value 184 evidenced as terminating both the Mendeleev and hydrational periodic systems. Such losses could be characteristic of natural or induced positive ionizations. Whether or not the human manipulation of imposed stresses will be able to bring about the formation of stable elements of neutral weights greater than 184 still appears to be a matter for the future to demonstrate and reveal.

The basic order which so repeatedly in preceding chapters has been represented as characterizing the ninety two naturally-occurring chemical elements of our earth may be appraised as an order applicable to the elements considered individually, apart from a common environment. In their associations they exhibit a wide range of relationships extending from complacent compatibility through confusion to combat and analogy with carnage. These elements are our composition and their relationship are our heritage. Archeologically and historically the rise and fall of civilizations has been linked intimately with contest and conquest. The tapping of nuclear energy has appreciably enhanced temptations and apprehensions. It is only through the selectivity involved in a very superior intelligence that humans may one day transcend their nature and create on earth the ability to enjoy it as a heaven far better than the varied and remote heavens born of ignorance and fear and so wistfully envisioned by so many for so long a time. That heaven on earth and for the living will be a continuing and supremely challenging goal for the intellect and its host in the age of science.

Nature during inconceivable numbers of centuries has evolved intricacies of atomic, molecular and electronic arrangement and interaction which to mankind's emergent mind seemed occult and supernaturally designed and manipulated. There need be no apology for initial ignorance and fear. Yet little by little increasing numbers of nature's intricacies have yielded to the relentless and insatiable curiosity that motivates science.

CHAPTER 20

HYDRATION AND THE ORGANIC WORLD

The context of the preceding chapter dealt with hydration and the inorganic world in a speculative manner without specific correlations to observational data. In the present chapter the considerations also will be speculative. It is to be recognized that the early differentiation of inorganic and organic has long since vanished. A precise differentiation of the inorganic and the organic in itself would constitute a challenging problem. In the present chapter it will suffice to let the word "organic" relate to substances commonly associated with organisms.

In Chapter 3 the observational data for the specific gravity of aqueous solutions—data for the most part obtained by German chemists in the closing decades of the previous century—were interpreted as evidencing an attraction between the involved solute ions and their solvent. This attraction in the anhydrous state was periodic and quantized in relation to ionic weight. The hydrational potentiality is projected as the fundamental attribute permitting and mediating the development of protoplasm and living organisms because it gave to inorganic units the flux essential to the elaboration of flexibility and elasticity, at the same time slowing their movement and restraining their ability to engage in chemical reactions.

In Chapter 9 analogous data for specific gravity were interpreted as clearly evidencing the bondaging potentialities conferred by the attractive force on the involved H_2O^- units in the absence of sufficient solvent to supply the prescribed amounts and provide for freedom of movement as independent hydrated ions in solution. Jointly the behavior patterns of solutes in aqueous solution and in the absence of adequate solvent were such as to implement and endow protoplasm as a suitable matrix for the seat of the metabolic processes of dissociation and synthesis supplying energy and substance for growth and development.

In Chapters 4 to 8 inclusive the observational data for the behavior patterns characterizing osmosis were interpreted as evidencing unexpected and important developments. Insofar as known, every living metabolically active cell possesses an ability to form osmotically-functional vacuoles. It is to be recognized that often what might be termed the impress of genetic factors so hastens differentiation as to make transitory or practically eliminate altogether the exercise of the ability to form vacuoles. Nevertheless the common incidence of vacuolar tissue in plants and the common inclusion of osmotically functional bladders in animals seems to characterize osmotic potentiality as an indispensable adjunct to organized protoplasmic metabolism.

It was recognized that the reciprocal inter-relationship of ionic hydrational potentiality on the one hand and the hydrational bonding of ions in the absence of free aqueous solvent on the other hand comprised insurance mechanisms suitable for the attainment and maintenance of the hydrostatic pressure so commonly present in living cells and so commonly designated as turgor. From the literature of physiological research one may venture that the development of a mathematical basis for the analysis of turgor—a basis implicit in the description of periodic hydrational potentiality in conjunction with osmosis—holds much of promise for clarification in the area.

It was recognized that the patterns of ionic behavior evidenced for osmosis in Chapters 4 and 5 supplied a basis for prospective progress in the analysis of absorption and translocation in plants and animals. The evidenced randomized exit of hydrated ions from osmotically active membranes supplied for multicellular organisms a mechanism operating in the direction of an equitable distribution of hydrated solutes among all cellular units. For individual roots in soil it supplied a mechanism accounting for the development of adjacent cylindrical regions rich with microorganisms and designated as rhizospheres. Collectively for plants in a common soil matrix the extrusion of hydrated solutes from root hair membranes supplied a mechanism operating in the direction of shared and therefore analogous resources and a consequent ecology.

The discovery of osmotic potentialities for the dissociation of solute radicals as documented in Chapter 7 opened up new avenues of inquiry in the broad area of nutrition. The existence of solute radicals in aqueous solution in reality had demonstrated the potentialities of water for dissociating chemical compounds, but it was made evident that osmosis involved the imposition of greater stresses and the allied or consequent breakdown of chemical bondages which commonly remained impervious to the dissociating potentialities of water. From the nutritional viewpoint it seemed of particular interest that sulfate, phosphate and carbonate radicals could be evidenced as having an intact status in aqueous solution and that in aqueous solutions containing these radicals they could be evidenced as subject to complete osmotic dissociation. This meant that in the vacuolated tissues of organisms osmosis could make available as protoplasmic metabolites elements which from a background of in vitro chemical behaviorism might be projected as chemically bound and hence not available. In the area of nutrition it was of interest further that under the special conditions indicated in Chapter 8 elemental nitrogen was added to the list of potential and prospective protoplasmic metabolites.

Speculatively in the previous chapter considerable attention was directed to the potential significance of the H^- ions evidenced in the data of Chapter 8 to the inorganic world. These ions appear to be of paramount significance in the organic world. Far more than any other atomic entity they embody the attribute of mobility while in the anhydrous state and excel in their readiness to release an electron on contact. Thus far more than any other atomic entity they become subject to appraisal as sources of energy.

Although the release of H^- ions as evidenced in Chapter 8 involved the dissociation of ammonium NH^+_4 radicals it was made apparent in the data of Chapter 17 that in the organic world H^- ions were particularly characteristic of cyclic sugars. This situation was interpreted as attributable to the fact that contrary to widespread contemporary appraisal the atoms entering into chemical compounds were always either negative or positive ions. As thus interpreted—and evidenced—amphoteric potentialities were restricted to aggregates of atoms, which might be quite polar, as illustrated by the acetate radical, or relatively non-polar, as represented by solute non-electrolytes. In cyclic carbon compounds the carbon ions were alternately positive and negative, a situation which greatly facilitated the incorporation of the extraordinarily dynamic H^- ions with their facility for releasing electrons.

It was recognized that phosphorylation in recent years had attained prominence as a projected adjunct process in the nutritional utilization of sugars. Yet the data of Chapters 7 and 15 had evidenced that the hydrogen ions associated with phosphate radicals were exclusively positive. The situation prompted the suggestion that the energy conventionally attributed to chemical bondages in phosphate radicals actually involved the release of electrons through interplay with hydrogen ions. A concrete characterization of energy seems precarious, but nutritionally the projected behavior patterns of electrons, of H^- ions, or even of $H°$ atoms appear preferable to chemical bondages as symbolic sources.

Correlated with the foregoing appraisal of cyclic sugars as sources of energy was a recognition of the data of Chapter 17 evidencing the existence of osmotic potentialities for the disruption of carbon-to-carbon linkages. The data were interpreted as having evidenced an incomplete dissociation of projected amphoteric solute ions. The researches conducted with a view to effecting a complete dissociation gave negative results. The respiratory loss of carbon dioxide from a matrix containing COH radicals might be projected as having mediated a release of ionic hydrogen, but an accompanying respiratory loss of water vapor remained to be integrated.

Conventionally for many years the operations involved in many metabolic processes have been attributed to real or fancied entities called enzymes. Whenever such explanations are valid they most certainly will not be impaired by further research. On the other hand the developments indicated in foregoing chapters suggest some reappraisals. One might venture, for example, that in the absence of a knowledge that osmosis possessed potentialities for the dissociation of certain solute radicals and for the disruption of carbon-to-carbon linkages it would be most natural to attribute such activities to the action of enzymes. The dissociation of carbohydrates is a subject of special interest in relation to metabolic respiratory processes sustaining life in plants and animals. These processes commonly are appraised as involving enzymes.

Oxygen as the dynamic primary sustainer of respiratory activity might be designated as the element most vital to the organic world. Quite naturally a great deal of attention has been directed to its role. In the classical researches of Pasteur *aerobic* signified contact with normal atmosphere and *anaerobic* signified a contrasting absence of such a contact. In the viewpoint of Pasteur, respiration in normal atmosphere had carbon dioxide as a characteristic by-product whereas respiration in the absence of oxygen had alcohol as a characteristic by-product. Although this latter respiration was abnormal for many organisms and for them was differentiated as fermentation, it was normal for some organisms and for them was differentiated as anaerobic respiration. Pasteur nevertheless recognized that a certain amount of anaerobic or fermentative respiration took place within animal tissues.

In general the area of atmospheric and protoplasmic contact in animals is the alveolar tissue of the lungs and in plants it is the alveolar tissue of leaf mesophyll. In animals the forcible contact is termed breathing. In plants an analogous forcible contact was reported by Bose in India but has remained unconfirmed.

If the word *aerobic* was to retain the significance occorded by Pasteur and be restricted to contact with air, aerobic respiration was a process confined to alveolar tissue. Then under what conditions did intracellular respiration involve the incorporation of carbon into alcohol which was retained, or involve the incorporation of carbon into carbon dioxide which was released ? For the maintenance of life in most organisms it appeared that intracellular respiration, however designated, had to be allied with alveolar respiration, which in turn involved oxygen. Oxygen therefore was to be appraised as a source of energy,—and energy was to be appraised as enigmatic.

Orientation to the behavior patterns of electrons has proceeded rapidly in recent years in the area of physics, as has been impressively evidenced by the developments documented in radio, television and electron microscopy. In the area of chemistry orientation has not been as rapid. Under the extraordinary stresses imposed incident to spectroscopy or mass spectrography the behavior patterns of electrons have been correlated with the behavior patterns of atomic ions, but at the lower levels of excitation which might be considered as more representative of the area of chemistry the role of electrons has received relatively little attention. If the role has been recognized as potentially important it has also been recognized as subtle and difficult to analyze.

With the development of the description of periodic hydrational potentiality it became obvious that the transfer of a single electron from one ion to another under common circumstances would modify the weights of both and hence modify their respective hydrational potentialities. Under specific circumstances, as at the transition points of the hydration periods, the loss of an electron would increase the hydrational potentiality from $1 H_2O^-$ unit to $23 H_2O^-$ units, or the gain of an electron would reverse the order of change. In recent years there has taken place a gradual increase in the recognition of the important role of hydration in chemical activity, but the recognition was made without reference to either the description of periodic hydrational potentiality or electrons.

It is of interest that within recent decades chemical science has included considerations of energy which projected for energy a most intimate association with chemical bonds. The bondages involved in phosphates were projected as particularly rich in energy. Yet in the data of Table 15 the chemical bondages in the phosphate radicals appeared to be readily subject to osmotic disruption in the manner indicated for various other radicals.

In a speculative way the contemplation of electrons as energy is most inviting. From a general background of physiology the identification of sucrose as a common source of energy is of interest in conjunction with the data of Chapter 17 because as represented in Diagram 4 sucrose contained six H⁻ ions per molecule. As represented in several chapters these H⁻ ions possessed very unusual characteristics among which were zero weight, exceedingly great mobility and a susceptibility to electron loss through contact. Thus a projected release of H⁻ ions from sucrose within a protoplasmic matrix would be followed by an immediate loss of electrons. Such losses would be subject to projection as occurring incident to the release of H⁻ ions from any organic compound, but the incidence of such ions in organic compounds might be projected as somewhat restricted, conditioned by the nature of the ions of carbon and oxygen involved. In simple carbon linkages, as in ethylene represented as $CH_2:CH_2$, two of the hydrogen atoms would bear negative charges—and responses to ethylene would be projected as related to a release of electrons by these ions. In the acetate radical the three hydrogen atoms would bear negative charges and would supply electrons following isolation and contact. In more complex organic compounds the involvement of H⁻ ions would be conditioned by the nature of the electrical charges on associated ions. The data of Chapters 6 and 8 suggested that in proteins the nitrogen bondages were greater than the carbon bondages but that nevertheless they were subject to disruption.

When the ultimate energy represented by organic compounds is projected as held by the H⁻ ion components the availability of the electrons within a protoplasmic matrix becomes suggested as conditioned not only by factors bringing about the release of the H⁻ ions from chemical bondage but also by the capabilities of associated ions to capture, retain and release the electrons freed from the H⁻ ions by contact. In such a capacity the potential ability of the evidenced P^{+5} ion might be noteworthy and account for the phosphate energy bonds of contemporary biochemistry.

In the pattern of these considerations the intracellular release of energy would be subject to appraisal as an anaerobic development with an ephemeral and modest measure of autonomy. Yet it is obvious that, as projected, the release of electrons would be accompanied by a build-up of hydrogen. Since multicellular plants and animals exposed to the normal atmospheric environment include aerobic tissues as necessary adjuncts, it seems to follow that oxygen must have some specific and dynamic role in relation to the projected intracellular protoplasmic metabolism. Quite fortunately precise knowledge in this area has not been essential.

A major objection to speculation is that it is without restraint : not that it is without interest. The primary role of oxygen could be projected as counteracting the imminent accumulation of hydrogen which otherwise would follow the release of energy. In such a capacity the intracellular product would be subject to projection as H_2O. The data of Chapter 17 suggested a common incidence of CHO units as protoplasmic metabolites, on which account a correlated role for oxygen could be projected as counteracting a potential accumulation of carbon through the formation of carbon dioxide. Both of these projected roles involve an allied translocation of oxygen from the alveolar tissue to other tissues, and both involve the concept of oxygen entering into chemical combination. Together they might supply an outline for the pattern of intracellular respiration,—a process which continues with progressive diminution after death. There is thus the suggestion that the aerobic metabolism in the alveolar tissue supplies not only oxygen but also the energy to implement its translocation. It was obvious that in animals such interludes as sleep and hibernation might bring modifications of the need for and use of such energy, but the fact that breathing was effected by involuntary muscles implied that it represented a process too vital to be endangered by absentmindedness. Did oxygen also have a role as a source of energy?

In the data of Table 22 in Chapter 12 it was to be noted that when attempts were made to impose positive electrical charges on oxygen, in experiments designed to permit the measurement of the relative mobilities of gaseous ions, the results obtained were such as to suggest that six electrons had been removed from the original oxygen molecules interpreted as neutral. In the same table it was to be noted that when attempts were made to impose positive electrical charges on air the results obtained were such as to suggest that only the oxygen of the air had become ionized and that six electrons had been removed from the neutral oxygen molecules. These data thus were in agreement in suggesting or evidencing that under the imposed conditions of stress each ionized molecule of oxygen gas gave off six electrons. As prescribed by the description of periodic hydrational potentiality such ions would have a weight of 44, a weight obviously identical with the weight of neutral carbon dioxide,—a situation which might or might not have any special significance.

It is common knowledge that solubility alone involves the operation of stresses sufficient to cause ionization, which is a redistribution of electrons in peripheral orbits. The data of Chapter 7 revealed that the stresses present within osmotically active membranes were greater than those involved in solubility, since they were able to bring about the dissociation of certain solute radicals. Thus for the conditions present in moist alveolar tissue one might project the presence of stresses sufficient to effect the release of up to six electrons per oxygen molecule without a disruption of the chemical bondage. Such a projection seemed supported by the results of experiments reported in Chapter 16, in which living plant tissues immersed in aqueous solutions of sulfates, selenates and tellurates gave off H_2S, H_2Se and H_2Te as analogs of the H_2O given off in normal respiration and as representing for the analogs of oxygen appreciable changes in the electron composition of the peripheral orbits. Such a projection represented aerobic respiration as a process effecting an immediate release of electrons as energy commonly functional in the translocation of oxygen for the mediation of intracellalar respiration. The release might be progressive and translocation might involve carriers.

The foregoing considerations of a speculative nature seemed of interest because at the atomic level they represented life on the earth as having been made possible by chance affiliations of atomic ions which permitted the capture, retentirn and release of electrons dispersed as H^- units of solar radiation. Yet periodic hydrational potentiality was represented as supplying both the bondage and flux for the elaboration of the organic world.

As thus viewed the seemingly wonderful accomplishments of nature and of modern science in the areas of both the inorganic and the organic have been based fundamentally on the same energy-electrons from the sun, The electrons have been operative through the same media—the ninety-two naturally-occurring chemical elements. In both the inorganic and the organic areas these elements have exhibited behavior patterns documenting a basic order commensurate with the simplicity of atomic number.

In a more general pattern of appraisal, electrons have been coming to our earth through seemingly infinite past ages. In the course of time transcendent adaptations to an ever-changing world have evolved. A dynamic incomprehensibly varied and intricate fabric of interrelationships has been woven from the elements representing basic order. Throughout eons mankind has been deeply involved in the adaptations and the weaving. From a satisfying basic order for atoms to a satisfying social order for mankind is not for out time, but the span marks a bright pathway for human potentialities continuously augmented through the unrelenting search for knowledge which is science.

*   *   *

Appendix A – Flint, L.H., *Journal of the Washington Academy of Science*
Vol. 22, No.5, March 4, 1932, 97-119:
"Hydration of the solute ions of the lighter elements."

JOURNAL

OF THE

WASHINGTON ACADEMY OF SCIENCES

Vol. 22   March 4, 1932   No. 5

CHEMISTRY.—*Hydration of the solute ions of the lighter elements.*[1]
L. H. Flint, Bureau of Plant Industry. (Communicated by G. N. Collins.)

Introduction

The researches carried out by Jones and his collaborators in the Chemical Laboratory of the Johns Hopkins University over a period of years and reported in various papers,[2] developed several independent lines of evidence, each of which pointed to the existence of hydrates in aqueous solutions. These researches may be said to constitute one of the fundamental bases of a relationship which in subsequent years has become widely recognized as an intimate one. At the present time the validity of a relationship between solute ions and their solvent is scarcely to be questioned.

The relationship between solute ions and solvent appears to be one of attraction, somewhat comparable with that characterizing many electrolytes crystallizing out of saturated solutions under certain conditions to form hydrated crystals. In the latter instance, however, a definite and usually integral number of water molecules is recognized as incorporated with each salt molecule. The relationship in both cases is termed hydration, but our knowledge of the specific molecular values involved in solutions is not sufficient to permit any precise evaluation corresponding with the use of the term as applied to crystals.

As a matter of fact we can find little satisfaction in our knowledge of the hydration of solute ions. We may observe that the velocity of a solute ion is not what we had expected it would be,—and may say that the ion is hydrated. We may note that a solute ion does not

---

[1] Received December 18, 1931.

[2] Amer. Chem. Journ. 23: 89, 1900; 31: 303, 1904; 33: 584, 1905; 37: 126, 1907. Carn. Inst. Wash. Pub. 60: 80, 1907; 180: 1913, and others.

depress the solubility of gases to the extent anticipated,—and may say that the ion is hydrated. We may find other real or apparent inconsistencies,—and attribute the results to the hydration of the ions. In resorting to the generalization we may often be correct,—but our explanation can scarcely become productive until our information is sufficient to permit mathematical expression.

Following a study of the hydration of ions over a period of several years, the writer has become convinced that a better understanding of this subject holds much of promise for the establishment of a more satisfactory interpretation of various inter-related solution phenomena. In outlining some of the reasons for such a conviction as a possible contribution to the subject it will be necessary to make two simple assumptions at the outset. These are (1) an inverse integral relationship between the anhydrous weight of a solute ion and the degree of its hydration, and (2) an orderly change in weight accompanying ionization. There are obvious objections to both these assumptions, and the objections may be sustained throughout the inquiry. Nevertheless, some of the suggested relationships which appear to follow the assumptions are of more than passing interest. A number of such relationships touching upon the characteristics of the solute ions of the lighter elements will be considered in this paper.

ELECTRICAL CONDUCTIVITY AS AN INDEX OF VELOCITY AND HYDRATION

Following the first assumption we may examine the observed electrical conductivities of simple solute element ions of the lighter elements and obtain the order of hydration indicated by the relative velocities of the ions. With univalent ions such as Na^+ and K^+ the conductivities and velocities may be considered to be of the same order, and through the extension of Graham's Law the relative velocity values of the ions Na^+ and K^+ (as derived from observed measurements of conductivity through the use of transference data) become indices of relative hydration. The assumption of hydration on an inverse integral basis requires that a succession of weight values be characterized by regular intervals. Such a requisite regularity does not. characterize the observed combining weights of the lighter elements, but is found in their atomic numbers. These numbers may be brought to the $O = 16$ scale by doubling, in which case a tentative series of regular weight values is attained as a basis upon which to project an assumed inverse integral hydration. The following conductance values for the ions Na^+ and K^+ at 18°C. are given by Nernst:[3] $K^+ = 65.3$, $Na^+ = 44.4$. These values, considered as relative velocities, permit the assumption of an inverse integral hydration only when the weight 38, representing potassium, has four water molecules attached, and the weight 22, representing sodium, has twelve water molecules attached. The weight, hydration and velocity values which

TABLE 1.—WEIGHT, HYDRATION AND VELOCITY VALUES CALCULATED FOR THE LIGHTER ELEMENTS THROUGH AN EXTENSION OF GAS LAWS IN RELATION TO OBSERVED ELECTRICAL CONDUCTIVITIES

A.N.	E	Assumed as W. $2 \times$ A.N.	V_1	Postulated Number of Water Molecules	Mol. Wt. Water of Hydration	Mol. Wt. Hydrated Molecule	V_2
0	—	0	—	23	414	414	491
1	H	2	7082	22	396	398	501
2	He	4	5000	21	378	382	512
3	Li	6	4090	20	360	366	523
4	Be	8	3546	19	342	350	535
5	B	10	3162	18	324	334	547
6	C	12	2887	17	306	318	561
7	N	14	2672	16	288	302	575
8	O	16	2500	15	270	286	591
9	F	18	2357	14	252	270	609
10	Ne	20	2236	13	234	254	627
11	Na	22	2132	12	216	238	648
12	Mg	24	2041	11	198	222	671
13	Al	26	1961	10	180	206	698
14	Si	28	1890	9	162	190	726
15	P	30	1826	8	144	174	758
16	S	32	1768	7	126	158	796
17	Cl	34	1715	6	108	142	839
18	A	36	1667	5	90	126	891
19	K	38	1622	4	72	110	953
20	Ca	40	1581	3	54	94	1031
21	Sc	42	1543	2	36	78	1133
22	Ti	44	1508	1	18	62	1271
23	V	46	1474	0	0	46	1474

thus develop for the lighter elements from the assumed inverse integral relationship in conjunction with observed electrical conductivities are brought together in Table 1.

In Table 1 the first column gives the atomic number of the element, the second column gives the chemical symbol of the element, and the third column gives the atomic number transposed to the familiar O = 16

[3] Citation on page 177 in Bayliss, W. M. *Principles of General Physiology*. 1915.

scale as an expression of weight. The fourth column gives the reciprocals of the square-roots of these weight values, multiplied by 10^4 for convenience in manipulation. These reciprocal values represent theoretical relative velocities as derived through the extension of Graham's Law under an assumption of *no hydration*, and on this account they have been designated as V_1 values. The fifth column gives the numbers of water molecules which must be postulated as characterizing hydration when an inverse integral relationship between weight and hydration is associated with the observed relative conductances of potassium and sodium considered as of weight 38 and 22 respectively. The sixth column gives the weight of these water molecules, and the seventh column the total weight of the elements represented as so hydrated. The eighth column gives the reciprocals of the square-roots of these "hydrated weight" values, multiplied by 10^4 for convenience in manipulation. The values of the eighth column represent the theoretical velocities under the indicated hydration as derived through the extension of Graham's Law, and have been designated as V_2 values.

The series as above tabulated comprises the elements of the first quarter of the periodic system,—a unique division. These elements are hereinafter arbitrarily termed the *lighter* elements as distinguished from the remaining heavier elements of the periodic system.

We may study the possible usefulness of Table 1 by using the second assumption of the paper in connection with it,—namely, the assumption that a regular change in weight accompanies ionization. Under such an assumption the most natural increment of change is that which would be effected by the gain or loss of a unit electrical charge on the nucleus of an atom, by virtue of which the weight characteristic represented in column three of Table 1 would be subject to unit change. There are objections to the assumption of a change in weight as an accompaniment of ionization. These objections do not appear to be as serious at present as they would have been before the advent of an electrical interpretation of matter and a knowledge of the modifications characterizing radioactive elements. Nevertheless, the objections to the second assumption may be sustained throughout the inquiry,— notwithstanding which it will be of interest to examine some of the relationships which appear to follow the assumption.

With specific reference to the potassium and sodium ions, K^+ and Na^+, it follows from the above assumption that the regular weight values assigned to the elements, potassium (38) and sodium (22), in the consideration of their hydration as indicated by relative velocity now become subject to further description as relative weights of the un-ionized or "neutral" elements. These weights as assigned to potassium and sodium further become subject to unit modification for unit charge characterizing the ionized state, and since the potassium and sodium ions being considered have a single positive charge each, it follows that the weights of the so-called "neutral" elements would advance one step upon becoming so ionized. The weight values representing the potassium and sodium ions thus become $K^+ = 40$ and $Na^+ = 24$, and the hydration characterizing these ions as derived from Table 1 is now to be noted as 3 H_2O with K^+ instead of 4 H_2O and 11 H_2O with Na^+ instead of 12 H_2O, the values 4 H_2O and 12 H_2O still representing the indicated degree of hydration characterizing ions of weight 38 and 22 respectively. We now have the K and Na ions with weights modified from the regular values tentatively assigned as prerequisites of an inverse integral hydration relationship and subsequently described as the relative weights of the un-ionized or "neutral" atoms. These modified weights may be further designated as the relative weights of the anhydrous ions, or as "ionic" weights. If the anhydrous ions thus characterized by weight hydrate to the degree corresponding to such a weight, as derived from observed conductivities and indicated in Table 1, it follows that we are now in a position to study the relationship which would have to follow such a hydration system. On the other hand, if the anhydrous ions thus characterized by weight do not hydrate at all, but remain as unhydrated solute ions, it follows that we are also in a position to study the relationships which would have to follow that system. In other words, although we are interesting ourselves primarily in a hydration relationship, we are nevertheless in a strategic position to note an absence of hydration, should any solute ions appear from other considerations to be so characterized.

Before taking up the examination of observed measurements in relation to the two fundamental assumptions of this paper and to the hydration system embodied in Table 1 it may be to our advantage to recapitulate with respect to the use of the word "weight." Our first assumption of an inverse integral hydration required a regular system of weight values. Since the observed *combining weights* did not afford such a system the atomic numbers were doubled to obtain a tentative series of weight values later designated as the relative weights of the un-ionized or "neutral" atoms. Our second assumption of an orderly change in weight accompanying ionization gave us new values designated as relative weights of the ions, or "ionic" weights. It is to be noted that neither the weights of the "neutral" atoms nor the weights of the ions correspond to the weights characterizing the atoms in combination and commonly designated as the combining weights of the atoms, or more simply, the atomic weights. The possible interrelations of the three designations of weight will be considered at a subsequent point.

We may now turn our attention to the study of the possible usefulness of Table 1 in the prediction of solution characteristics. It immediately becomes evident that in the event the indicated numbers of water molecules combine with the respective element-ions to form hydrated ions, the transfer of such water from the solvent to the solute must profoundly influence the concentration of ions thus hydrated. Jones and his co-workers recognized this fact, but they were without a tentative basis for evaluating the extent of the influence. Table 1 affords such a basis.

The molecular weight values of the hydrated ions of the lighter elements may be readily derived from column seven of Table 1. For example, the molecular weight of the ion K^+ when hydrated may be derived as follows:

$$K = 38, K^+ = 38 + 2 = 40, K^+ \text{ hydrated} = K^+ + 3\,H_2O = K^+ +$$
$$(3 \times 18) = 94$$

From the summation weights of the ions characterizing a solution of any electrolyte comprising such ions, the extent of the influence of the assumed hydration may be mathematically calculated. For example, the relative amounts of solvent and solute characterizing a 1.0 molecular solution of KCl would be derived as follows:

$K = 38, K^+ = 40, K^+ \text{ hydrated} = K^+ (40 \text{ gms.}) + 3\,H_2O$ (54 gms.) = hydrated K^+, 94 gms.

$Cl = 34, Cl^- = 32, Cl^- \text{ hydrated} = Cl^-$ (32 gms.) $+ 7\,H_2O$ (126 gms.) = hydrated Cl^-, 158 gms.

94 gms. hydrated K^+ + 158 gms. hydrated Cl^- = 252 gms. solute.

In a solution 1.0 molecular made up to 1000 grams, the amount of solvent present would be calculated as $1000 - 252 = 748$ gms., or 74.8% of the amount present at "zero" concentration of solute. In a solution 1.0 molecular made up to a liter with observed combining weights the concentration would be 1.035 on the above basis by virtue of an observed weight of 74.553 as compared with a calculation weight of 72. Moreover, the total weight of solution under the observed conditions would not be 1000 grams, since the density of the solution is not that of the solvent. Yet a relationship between combining weights and the assumed weights for "neutral" and "ionized" atoms can not be considered in this paper without involving argument in digression. Furthermore, in some electrical conductivities of concentrated solutions as observed by various investigators the values have been transposed from a volume to a weight basis through "corrections."[4] Under the circumstances in a reconnaissance survey of certain relationships which appear to follow our initial assumptions we may disregard the factors which differentiate the two bases, and entertain a degree of tolerance for approximate agreements.

Through the use of Table 1, then, we have calculated that at 1.0 molecular concentration a solution of KCl contains 74.8% of the weight of solvent which characterizes it at "zero" concentration. The observed specific molecular conductivities of KCl at "zero" and 1.0 molecular concentrations, 18°C., as given by Noyes and Falk,[5] are 130.0 and 96.5 respectively. The relative specific molecular conductivity is thus $96.5 \div 130.0 = .742$, or 74.2%.

The obvious suggestion following the order of agreement noted is that the decrease in specific molecular conductivity with concentration, which is quite generally interpreted as indicating incomplete dissociation of the electrolyte, may in reality be an index of the relative weight of solvent present in a solution of a completely ionized electrolyte.

Yet with respect to observed conductivity measurements such an interpretation leads to the inference that the values for concentration, ranging from 1.0 molecular to "zero" molecular, for example, would involve the use of varying bases. If such should prove to be the case the order of relative specific molecular conductivities of various electrolytes comprising ions of the lighter elements should be predictable from the summations of velocities as given in Table 1, through modification to the extent indicated by the summations of weight values (also given in Table 1) in relation to the amount of solvent present. Thus, the amount of solvent characterizing a 1.0 molecular solution of KCl was calculated above from Table 1, as 74.8%. The summation

---

[4] For example, see Noyes and Falk. Journ. Amer. Chem. Soc., **34**: 454, 1912.
[5] Previous citation.

TABLE 2.—COMPARISON OF OBSERVED SPECIFIC MOLECULAR ELECTRICAL CONDUCTIVITIES OF SOME ELECTROLYTES INVOLVING IONS OF THE LIGHTER ELEMENTS WITH VELOCITY, HYDRATION AND WEIGHT CONSIDERATIONS EMBODIED IN TABLE 1.

Electrolyte	Observed Weight of Variant Ion of Series	Assumed Weight of Anhydration of Ions (From Table 1)	Assumed Hydration of Ions (From Table 1)	Assumed Weight of Hydrated Ions (From Table 1)	Calculation for % Solvent at 1.0 mol. Concentration	Assumed Relative Velocity of Hydrated Ions (From Table 1)	Calculation for unit molecular Velocity	Calculated Relative unit Velocity at 1.0 mol. conc.	Observed Relative Unit sp. Mol. Cond. at 1.0 mol. conc.	Calculated Relative % Solvent present at 1.0 mol. conc.	Observed Relative Sp. Mol. Conductivity at 1.0 mol. conc.
LiCl	6.94	Li+ = 8 (comb. wt.), Cl− = 7 (comb. wt.)	Li+ = 19 H₂O, Cl− = 7 H₂O, Total = 26 H₂O	Li+ = 350, Cl− = 158, Total = 508	508 − 40 = 468, 1000 − 468 = 532, 532 ÷ 1000, .532, or 53.2%	Li+ = 535, Cl− = 796, Total = 1331	.532 × 1331 = 708	708	762	53.2	60.7
NaCl	22.997	Na+ = 24 (comb. wt.), Cl− = 23 (comb. wt.)	Na+ = 11 H₂O, Cl− = 7 H₂O, Total = 18 H₂O	Na+ = 222, Cl− = 158, Total = 380	380 − 56 = 324, 1000 − 324 = 676, 676 ÷ 1000, .676, or 67.6%	Na+ = 671, Cl− = 796, Total = 1467	.676 × 1467 = 992	992	988	67.6	67.8
MgCl₂	24.32	Mg++ = 28 (comb. wt.), Cl− = 26 (comb. wt.), Cl− = 33 (comb. wt.)	Mg++ = 9 H₂O, Cl− = 7 H₂O, Cl− = 7 H₂O, Total = 23 H₂O	Mg++ = 190, Cl− = 158, Cl− = 158, Total = 506	1000 − 506 = 494, .494, or 49.4%	Mg++ = 726, Cl− = 796, Cl− = 796, Total = 1522	506 ÷ 2 = 253, 1000 − 253 = 747, 747 ÷ 1000 = .747, .747 × 1522 = 1137.5	1137.5	1254	49.4	48.7
AlCl₃	26.97	Al+++ = 32 (comb. wt.), Cl− = 29 (comb. wt.), Cl− = 32 (comb. wt.), Cl− = 33 (comb. wt.)	Al+++ = 7 H₂O, Cl− = 7 H₂O, Cl− = 7 H₂O, Cl− = 7 H₂O, Total = 28 H₂O	Al+++ = 158, Cl− = 158, Cl− = 158, Cl− = 158, Total = 632	1000 − 632 = 368, .368, or 36.8%	Al+++ = 796, Cl− = 796, Total = 1592	632 ÷ 3 = 210.66, 1000 − 210.66 = 789.3, 789.3 ÷ 1000 = .789, .789 × 1592 = 1256	1256	1288	36.8	36.4
KCl	39.096	K+ = 40 (comb. wt.), Cl− = 33 (comb. wt.)	K+ = 3 H₂O, Cl− = 7 H₂O, Total = 10 H₂O	K+ = 94, Cl− = 158, Total = 252	1000 − 252 = 748, .748, or 74.8%	K+ = 1031, Cl− = 796, Total = 1827	.748 × 1827 = 1367	1367	1367 (base)	74.8	74.2
CaCl₂	40.07	Ca++ = 44 (comb. wt.), Cl− = 42 (comb. wt.), Cl− = 33 (comb. wt.)	Ca++ = 1 H₂O, Cl− = 7 H₂O, Cl− = 7 H₂O, Total = 15 H₂O	Ca++ = 62, Cl− = 158, Cl− = 158, Total = 378	1000 − 378 = 622, .622, or 62.2%	Ca++ = 1271, Cl− = 796, Cl− = 796, Total = 2067	378 ÷ 2 = 189, 1000 − 189 = 811, 811 ÷ 1000 = .811, .811 × 2067 = 1675	1675	1570	62.2	61.2
KF	19.0	K+ = 40 (comb. wt.), F− = 16 (comb. wt.)	K+ = 3 H₂O, F− = 15 H₂O, Total = 18 H₂O	K+ = 94, F− = 286, Total = 380	380 − 56 = 324, 1000 − 324 = 676, 676 ÷ 1000 = .676, or 67.6%	K+ = 1031, F− = 591, Total = 1622	.676 × 1622 = 1097	1097	1057	67.6	68.6
KCl	35.457	K+ = 40 (comb. wt.), Cl− = 32 (comb. wt.), Cl− = 33 (comb. wt.)	K+ = 3 H₂O, Cl− = 7 H₂O, Total = 10 H₂O	K+ = 40, Cl− = 158, Total = 252	1000 − 252 = 748, .748, or 74.8%	K+ = 1031, Cl− = 796, Total = 1827	.748 × 1827 = 1367	1367	1367 (base)	74.8	75.5

of the velocity values corresponding to the molecular weights of the hydrated ions K and Cl may be derived from Table 1, column 8, as follows:

$$K = 38, \; K^+ = 40, \; K^+ \text{ hydrated} = 40 + 3 \; H_2O, \; V_2 = 1031$$
$$Cl = 34, \; Cl^- = 32, \; Cl^- \text{ hydrated} = 32 + 7 \; H_2O, \; V_2 = 796$$

Summation = 1827

If, now, we represent the specific molecular conductivity of KCl at "zero" concentration by the summation velocity value 1827, at 1.0 molecular concentration the apparent relative velocity value may be calculated as 74.8% of 1827, or 1367, since the transfer of water from solvent to solute under the assumed inverse integral hydration system would reduce the apparent concentration of solvent, as previously noted.

We may now examine a group of electrolytes involving ions of the lighter elements with respect to the above suggestions and the assumed interrelations of weight, hydration and velocity embodied in Table 1. To facilitate such an examination the respective data are brought together in Table 2.

In Table 2, the electrolytes in the respective series are arranged in the order of the increasing atomic weight of the variant ion, as indicated in the second column. In such series the assumption of an inverse integral relationship between weight and hydration would yield relative velocities of an order increasing with weight as calculated

---

[*] Observed values may involve weight-normal or volume-normal solutions. The observed values cited for LiCl, NaCl and KF involved no "correction" to weight-normal basis, and calculations are, therefore, made to correspond with the observed volume-normal basis. The other observed values involve "corrections" to the weight-normal basis. All observed values in the first series are at 0°C., and in second series at 18°C., except as otherwise noted.

[†] LiCl, Jones and Getman, Ztschr. phys. Chem. 46, 1903, p. 262, 1.67 m = 42.28, .88 m = 35.58, by interpolation 1.0 mol. = 36.597; NaCl, Int. Crit. Tables, Vol. VI, p. 233, 1.0 mol. = 47.5; MgCl$_2$, Jones, Carn. Inst. Wash. Pub. 180, p. 65, .9415 m = 60.31; AlCl$_3$ same p. 78, 1.0 mol. = 61.93; KCl, same p. 16, 1.05 mol. = 65.7; CaCl$_2$ same, p. 16, 1.0 mol. = 75.5; KF, Noyes and Falk, J. A. Chem. Soc. 34, 1912, p. 463, 1.0 mol. = 75.95; KCl, same, p. 463, 1.0 mol. = 98.22.

[‡] Additional values as follows: LiCl, Washburn, J. A. Chem. Soc. 33, 1911, p. 1474, "zero" conc. = 60.3; NaCl, Kohlrausch and Holborn, Leitvermögen der Elektrolyte, 1898, p. 158 [18°C.], 1.0 mol. = 74.4, .0001 mol. = 109.7; MgCl$_2$, Jones, Carn. Inst. Wash. Pub. 180, p. 65, "zero" conc. = 123.95; AlCl$_3$, same, p. 78, "zero" conc. = 170; KCl, Noyes and Falk [18°C.], J. A. Chem. Soc. 34, 1912, p. 461, "zero" conc. = 130.0, p. 462, 1.0 mol. conc. = 96.5; CaCl$_2$, Jones, Carn. Inst. Wash. Pub. 180, p. 63, "zero" conc. = 123.46; KF, Noyes and Falk, J. Am. Chem. Soc. 34, 1912, p. 461, "zero" conc. = 111.2.

from Table 1 and given in the ninth column. The observed velocities as inferred from the relative unit specific molecular conductivities given in the tenth column are to be noted as of corresponding order. On the other hand, the specific hydration values derived from the above assumption as related to the observed relative ion conductance of the Na and K ions yield an entirely random series of values for the amount of solvent characterizing 1.0 molecular solutions of the respective electrolytes as derived in the fifth and sixth columns and given in the seventh column. The observed relative specific molecular conductivities at 1.0 molecular concentration given in the twelfth column are to be noted as of a corresponding order. In the last four columns of Table 2 we have, therefore, a double-checking series of comparisons relating the values of Table 1 to observed measurements. On the one hand relative velocities predicted upon the assumed hydration through the extension of Graham's Law are to be noted as in substantial agreement with observed relative conductivities. On the other hand, the changes in concentration of solvent which would be anticipated from the assumed hydration are to be noted as in agreement with the observed apparent modifications of specific molecular conductivities. These interlocking series of comparisons involving two aspects of solution phenomena as measured by electrical conductivity thus appear to be characterized by agreements beyond the possibility of mere accident.

In the third column of Table 2, it may be noted that assumed combining weights are given in parentheses below the assumed weights of the anhydrous ions. These combining weights are intermediate between the respective "neutral" and "ionic" weights previously discussed, and mathematically represent the sharing of electrons in chemical combination, considered from the standpoint of weight-change as an accompaniment of ionization. For example, if the element sodium represented as Na is assigned a weight of 22 in the un-ionized or "neutral" state, and a weight of 24 following the loss of an electron to become positively ionized as Na^+, then when it shares a single electron in combination, its combining weight quite naturally may be assumed as the intermediate value, or 23. The values thereby attained as combining weights for such ions as Li^+, Na^+ and K^+ do not depart appreciably from observed values, but since the corresponding values for many other ions are at variance with observed values the matter of combining weights of the lighter elements requires particular consideration in relation to the assumptions regarding hydration and weight change. If we are to make the two initial assumptions of this paper it appears incumbent upon us to eventually define combining weight and relate it to a substantial group of observed measurements. At this time, however, the matter of the hydration of the ions must take precedence, and their combining weights must remain parenthetical.

We may now continue the inquiry through an examination of the measurements of other phases of solution phenomena.

Freezing-Point Depression as an Index of Hydration in the Lighter Element Ions.

The work of Raoult[1] established an importance for the freezing-point depression of a solvent effected by a solute, and in subsequent years the measurement has become important in the determination of the molecular weights of dissolved non-electrolytes, in the measurement of osmotic pressure of both electrolytes and non-electrolytes and in the measurement of the electrolytic dissociation of electrolytes. As related to each of these applications the usefulness of the measurement appears to be largely restricted to dilute solutions. As related to osmotic pressure the measurement has attained importance through analogy, following the demonstration of a direct proportionality between freezing-point lowering and osmotic pressure. As related to the electrolytic dissociation of electrolytes the measurement attained importance through inference, following the interpretation of the decrease in specific molecular conductivity with increase in concentration as an index of incomplete dissociation.

Yet in the foregoing section observed electrical conductivity values for solutions involving ions of the lighter elements were noted as corresponding with hydration, weight and velocity values predictable from Table 1. The apparent decrease in observed specific molecular electrical conductivity with increase in concentration was suggested as associated with unevaluated changes in concentration caused by the hydration of the ions. The "true" specific molecular conductivity was thus indicated as a constant, whereupon complete ionization at all concentrations became characteristic of all solutions under the initial assumptions of the paper. It will be of interest to study the freezing-point depression of electrolytes involving the lighter element-ions with respect to the considerations embodied in Table 1.

In dilute solutions the depression of the freezing-point of the solvent by the solute is considered proportional to the number of molecules or ions of solute present.

[1] Ann. Chem. Phys. 28: 137, 1883; 2, 66, 1884.

We may now examine some observed freezing-point depressions with particular reference to the familiar relationship underlying the interpretation of freezing-point data, namely, that the gram molecular weight of a non-ionizing solute added to 1000 gms. of water reduces the freezing-point by 1.86°C. This relationship is considered as subject to direct modification through ionization, a solute giving rise to two ions, as KCl, effecting a reduction of 2 × 1.86°C., or 3.72°C. and a solute giving rise to three ions, as $CaCl_2$, effecting a reduction of 3 × 1.86°C., or 5.58°C. The degree of agreement between the values postulated under such a relationship and the observed values is commonly interpreted as a measure of dissociation. Yet under the assumptions of this paper electrical conductivities suggest complete ionization at concentrations up to 1.0 molecular KCl or its ionic equivalent.

TWO-ION ELECTROLYTES

Lithium Chloride, LiCl: The following observed values for the freezing-point depression of this electrolyte may be cited:[10] 1.0 mol. = 3.80°; .7939 mol. = 2.945°; .5012 mol. = 1.81°; .2474 = 0.86°. The summation weight representing the hydrated solute LiCl at 1.0 mol. concentration as derived from Table 1 is 508, and the proportionate values of the above concentrations may be calculated as follows: 1.0 mol. = 1 × 508 = 508; .7939 mol. = .7939 × 508 = 403.5; .5012 mol. = .5012 × 508 = 254.7; .2474 mol. = .2474 × 508 = 125.8. The observed freezing-point depressions are plotted against these proportionate values in the graph shown in Figure 1 and connected by a heavy line. For comparison we may venture to indicate the freezing-point depression of LiCl when the complete ionization suggested by electrical conductivity measurements is assumed. This depression would be 2 × 1.86°C. (unit molecular depression), or 3.72°C. for a summation weight of an electrolyte forming two ions. The weight value, 508, represents the amount of solute at 1.0 molecular concentration. The values are represented in the graph shown in Figure 1 as a broken line.

Sodium Chloride, NaCl. The following observed values for the freezing-point depression of this electrolyte may be cited:[11] 1.0 mol. = 3.37°; .700 mol. = 2.4°; .4293 mol. = 1.447°; .2325 mol. = .796°.

[10] 1.0 mol. from Int. Crit. Tables, 4: 258. Other values from page 227 in Smithsonian Tables, 6th Edition, 1914.

[11] 1.0 mol. from Int. Crit. Tables. 4: 258. Other values from page 227 in Smithsonian Tables, 6th Edition, 1914.

The summation weight representing the hydrated solute, NaCl, at 1.0 mol. concentration as derived from Table 1 is 380, and the proportionate values of the above concentrations may be calculated as follows: 1.0 mol. = 1 × 380 = 380; .700 mol. = .700 × 380 = 266; .4293 mol. = .4293 × 380 = 163.2; .2325 mol. = .2325 × 380 = 88.4. The observed freezing-point depressions are plotted against these proportionate values in the graph shown in Figure 1 and connected by a heavy line. The calculated depression for the ionization Na+

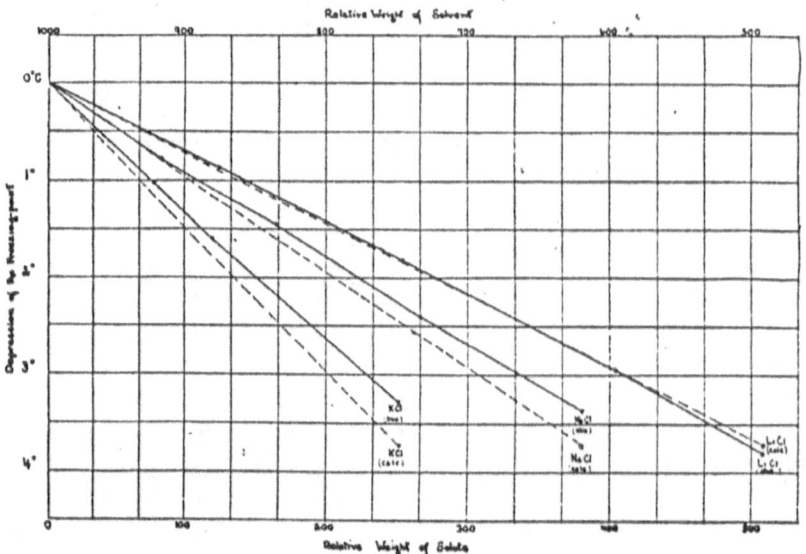

Fig. 1. Observed and calculated freezing-point depressions for some two-ion electrolytes.

and Cl⁻ would be 2 × 1.86°C., or 3.72°C. for a summation weight of 380, representing the amount of solute present at 1.0 molecular concentration. The values are represented in the graph shown in Figure 2 as a broken line.

Potassium Chloride, KCl. The following observed values for the freezing-point depression of this electrolyte may be cited:[12] 1.0 mol. = 3.268°; .476 mol. = 1.605; .3139 mol. = 1.07. The summation weight representing the hydrated solute KCl at 1.0 mol. concentration as derived from Table 1 is 252, and the proportionate values of the above concentrations may be calculated as follows: 1.0 mol. = 1 × 252 = 252; .476 mol. = .476 × 252 = 120; .3139 mol. = .3139 ×

[12] Smithsonian Tables, 6th Edition, p. 227, 1914.

252 = 79.2. The observed freezing-point depressions are plotted against these proportionate values in the graph shown in Figure 1 and connected by a heavy line. The calculated depression for the ionization K^+ and Cl^- would be $2 \times 1.86°C.$, or $3.72°C.$ for a summation weight of 252, representing the amount of solute present at 1.0 molecular concentration. The values are represented in the graph shown in Figure 1 as a dotted line.

The comparisons set forth in the graph shown in Figure 1 indicate only a general order of agreement, yet they appear to warrant the

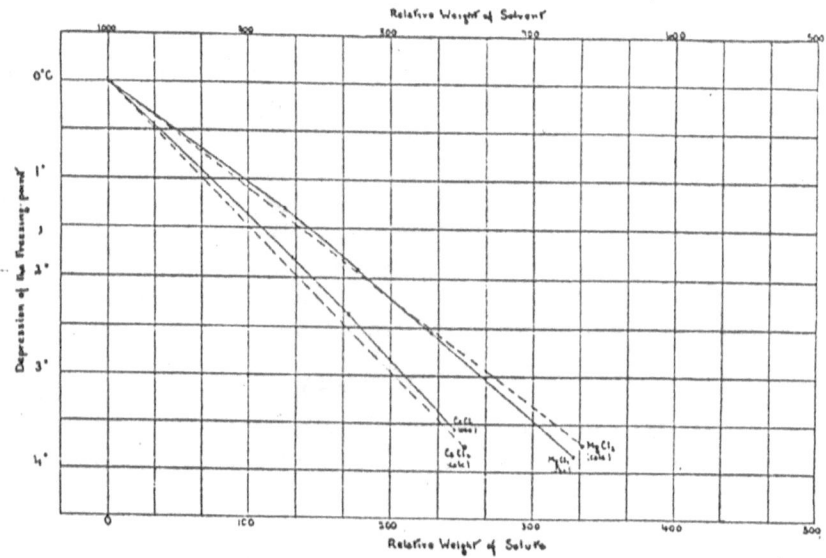

Fig. 2. Observed and calculated freezing-point depressions for some three-ion electrolytes.

consideration of other electrolytes at corresponding ionic concentrations.

THREE-ION ELECTROLYTES

Magnesium Chloride, MgCl₂. The following observed values for the freezing-point depression of this electrolyte may be cited.[11]   .65 mol. = 3.854; .45 mol. = 2.537; .35 mol. = 1.910; .25 mol. = 1.306. The summation weight representing the hydrated solute $MgCl_2$ at 1.0 molecular concentration as derived from Table 1 is 506, and the proportionate values of the above concentrations may be calculated

[11] Jones, H. C. Carn. Inst. Wash. Pub. 180: 23, 1913.

as follows: .65 mol. = .65 × 506 = 329; .45 mol. = .45 × 506 = 228; .35 mol. = .35 × 506 = 177; .25 mol. = .25 × 506 = 126.5. The observed freezing-point depressions are plotted against these proportionate values in the graph shown in Figure 2 and connected by a heavy line. At .666 mol. concentration, equi-ionic with 1.0 molecular KCl, the calculated freezing-point depression would be $\frac{2}{3}$ × 3 × 1.86°C. = 3.72°C., for a summation weight of .666 × 506, or 337. These values are represented in the graph as a broken line.

Calcium Chloride, CaCl₂. The following observed values for the freezing-point depression of this electrolyte may be cited:[14] .65 mol. = 3.55°; .45 mol. = 2.35°; .35 mol. = 1.801°. The summation weight representing the hydrated solute, $CaCl_2$, at 1.0 molecular concentration as derived from Table 1 is 378, and the proportionate values of the above concentrations may be calculated as follows: .65 mol. = .65 × 378 = 245.8; .45 mol. = .45 × 378 = 170; .35 mol. = .35 × 378 = 132.3. The observed freezing-point depressions are plotted against these proportionate values in the graph in Figure 2 and connected by a heavy line. At .666 mol. concentration the suggested freezing-point depression would be $\frac{2}{3}$ × 3 × 1.86°C., or 3.72°C., for a summation weight of .666 × 378, or 252. These values are represented in the graph shown in Figure 2 as a broken line.

FOUR-ION ELECTROLYTE

Aluminum Chloride, AlCl₃. The following observed values for the freezing-point depression of this electrolyte may be cited:[15] .50 mol. = 3.9446°; .4 mol. = 2.910; .25 mol. = 1.6604°; .2 mol. = 1.279. The summation weight representing the hydrated solute $AlCl_3$ at 1.0 molecular concentration as derived from Table 1 is 632, and the proportionate values of the above concentrations may be calculated as follows: .50 mol. = .50 × 632 = 316; .4 mol. = .4 × 632 = 252.8; .25 mol. = .25 × 632 = 158; .2 mol. = .2 × 632 = 126.4. The observed freezing-point depressions are plotted against these proportionate values in the graph shown in Figure 3 and connected by a heavy line. At .50 molecular concentration, equi-ionic with 1.0 molecular KCl, the calculated freezing-point depression would be $\frac{1}{2}$ × 4 × 1.86°C., or 3.72°, for a summation weight of $\frac{1}{2}$ × 632, or 316. These values are represented in the same graph as a dotted line.

The order of agreement to be noted in the foregoing graphs indicates

---

[14] Jones, H. C. Carn. Inst. Wash. Pub. 180: 22, 1913.
[15] Jones, H. C. Carn. Inst. Wash. Pub. 180: 78, 46, 1913.

that the suggestion of complete ionization at concentrations as great as 1.0 molecular, a suggestion arising from a consideration of electrical conductivity measurements, is not without support in the data of freezing-point determinations.   It is readily apparent that at the lower concentrations shown in the graphs the observed freezing-point depressions are less than the calculated values.   These differences are substantially off-set when the concentrations of the observed values are recalculated on the weight-normal basis used in apportioning the relative amounts of solute and solvent.   It is further apparent from

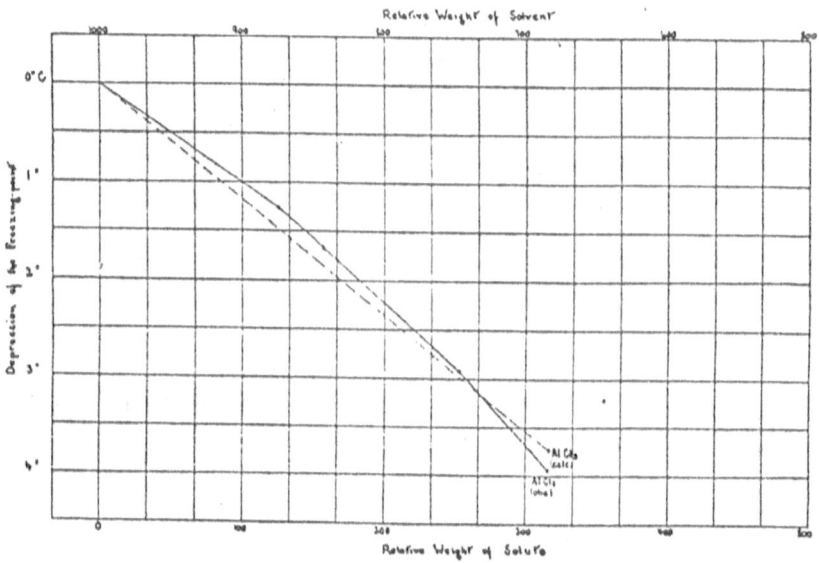

Fig. 3. Observed and calculated freezing-point depressions for a four-ion electrolyte

these graphs, moreover, that at concentrations approaching one molecular (for KCl type), or its ionic equivalent, the observed freezing-point depressions of such electrolytes as LiCl, MgCl$_2$ and AlCl$_3$ are greater than those anticipated under the calculated straight-line relationship.   These electrolytes are indicated in Table 2 as being characterized by the greater degree of hydration, suggested as a significant factor in the more concentrated solutions.   It appears to be of interest, therefore, to examine the freezing-point measurement to the point of solidification of the solution.

A series of freezing-point measurements of solutions of CaCl$_2$ at various concentrations approaching the point at which the solution

204

solidifies are available from Jones,[16] and on this account this electrolyte will be examined in some detail with respect to hydration.

As derived from Table 1 the anhydrous calcium ion, Ca^{++}, has a weight of 44 (suggesting thereby a combining weight of 42), with a hydration of one water molecule, by virtue of which hydration the hydrated ion has a weight of 62. Similarly, the anhydrous chlorine ion, Cl^-, has a weight of 32 (suggesting thereby a combining weight of 33) with a hydration of seven water molecules, the weight of the hydrated ion being 158. On such a basis, therefore, the summation weights associated with $CaCl_2$ are as follows: anhydrous state, Ca^{++} = 44, Cl^- = 32, Cl^- = 32, total = 108; hydrated state, Ca^{++} = 62, Cl^- = 158, Cl^- = 158, total = 378. We may now calculate a series of characteristics for solutions of $CaCl_2$ at various concentrations on the foregoing basis. Taking 1000 grams of solution as a concentration standard, we may calculate the expected saturation point as follows: $1000 \div 378 = 2.645$. At 2.645 molecular on the weight basis the solution should be saturated, and when concentration is expressed as a weight of electrolyte added to 1000 gms. of water, (which is the basis of concentration used in freezing-point depression studies), and the observed weight of $CaCl_2$ is taken as 111, the 2.645 molecular value becomes 3.605 molecular, or 400 grams $CaCl_2$ added to 1000 gms. H_2O. The expected depression of the freezing-point at 2.645 molecular, assuming complete ionization, may be calculated as follows: $2.645 \times 3 \times 1.86° = 14.76°$. But whereas at zero concentration of solute there are 1000 grams of free solvent and anhydrous solute present, at 2.645 molecular concentration of solute there is no free solvent present and 286 gms. of solute. If the freezing-point depression is assumed to have been influenced by this change, the extent of the influence becomes measurable by simple division, in which the relative amount of solvent and anhydrous solute is expressed as a fraction. Thus $14.76° \div .286 = 51.6°$. On such a basis, therefore, we may calculate the expected additional depression of the freezing-point attributable to the transfer of water from solvent to solute under the assumed hydration. A series of values for $CaCl_2$ at various concentrations has been calculated and incorporated in Table 3, wherein is also cited a series of corresponding values from observed freezing-point depressions at various concentrations as obtained by Jones.

[16] Jones, H. C. Carn. Inst. Wash. Pub. **180**: 15, 1913.

We may now plot the values of Table 3 relating to proportion of solute and freezing-point depression. In the graph in Figure 4 calculated values appear represented by heavy lines. The straight line is derived from column four of the calculated series, and is an expression of the relationship fundamental to the interpretation of freezing-point depression—1.86°C. depression for each gram ion

TABLE 3.—DATA OF CALCULATED AND OBSERVED FREEZING-POINT DEPRESSION IN RELATION TO AN ASSUMED HYDRATION

Mol. Conc. (Observed Basis)	Grams Hydrated Solute	Conc. Hydrated Solute (Weight Basis)	Calc. Freezing-Point Depression (Conc. Hyd. Sol. × 3 × 1.86°)	Grams Free Solvent	Grams Anhydrous Solute	Grams Solvent Plus Grams Anhydrous Solute	Freezing-Point Depression Calculated for conc. on Observed Basis.	
			CALCULATED SERIES					
0	0	0	0°	1000	0	1000.	0°	
.5	184	.487	2.72°	816	52.6	868.6	3.13°	
1.0	350	.926	5.165°	650	100.	750.	6.89°	
1.5	500	1.323	7.38°	500	142.75	642.75	11.48°	
2.0	636	1.6825	9.385°	364	181.8	545.8	17.22°	
2.5	760	2.01	11.22°	240	217.3	457.3	24.52°	
3.0	876	2.318	12.93°	124	250.	374.	34.58°	
3.5	980	2.592	14.475°	20	280.	300.	48.25°	
3.605	1000	2.645	14.76°	0	286.	286.	51.6°	
			OBSERVED SERIES					
								Obs. Δ
.3	113	.299	1.67°	887	32.3	919.3	1.82°	1.517°
.7	253	.67	3.74°	747	72.2	819.2	4.57°	4.065°
1.0	350	.926	5.16°	650	100.	750.	6.88°	6.41°
1.4	471	1.246	6.95°	529	134.5	663.5	10.48°	10.05°
1.75	569	1.505	8.40°	431	162.6	593.6	14.17°	14.33°
2.2	687	1.8175	10.15°	313	196.3	509.3	19.95°	21.07°
2.7	808	2.138	11.92°	192	232.	424.	28.1°	30.25°
3.1	896	2.37	13.23°	104	256.	360.	36.75°	39.5°
3.51	980	2.59	14.45°	20	280.	300.	48.15°	49.5°

present—although the concentration basis has been modified from "a weight of electrolyte added to 1000 gms. water" to "a weight of electrolyte in 1000 gms. solution." This fundamental relationship thus involves the number of ions present, and in so doing further involves the implication that all solute ions are of the same size—an implication which also follows from the extension of the gas laws in relation to velocity as interpreted through observed electrical conductivities. The curved line derived from column eight of the same series is an expression of the modification of the straight-line relationship which might be anticipated as a result of the assumed hydration. In other words, under the assumed hydration each molecule of $CaCl_2$ removes fifteen water molecules of solvent, and the unevaluated concentration thereby brought about gives an apparent falling-off in freezing-point depression represented by the departure of the curved line from the straight line.

We may now examine the values for freezing-point depression as derived from column nine of the observed series and represented by a dotted line curve in the graph.

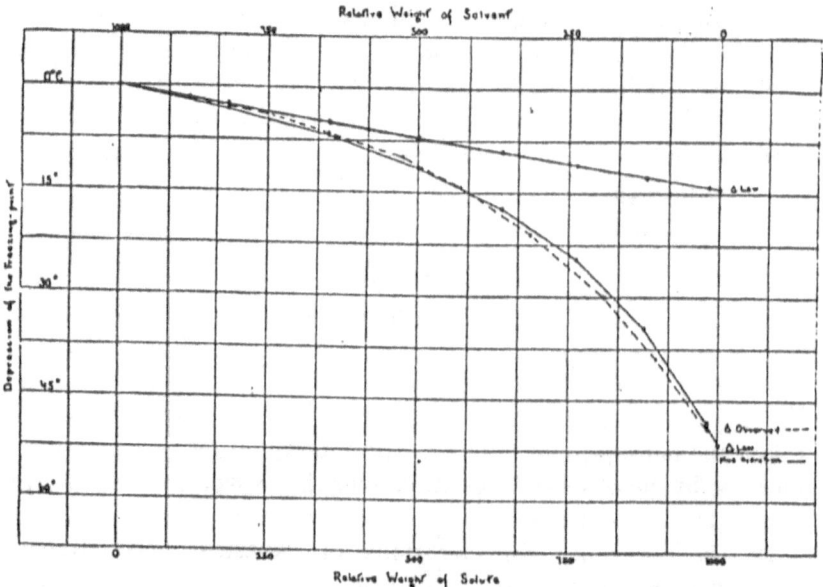

Fig. 4. Observed and calculated freezing-point depressions for aqueous solutions of $CaCl_2$ at all concentrations.

The agreement between the observed and calculated values as represented respectively by the dotted and full curved lines appears to be beyond the possibility of accident. The calculated point at which the mixture of $CaCl_2$ and water becomes 100% hydrated $CaCl_2$ ($-51.6°C.$) corresponds with the observed cryohydric or eutectic point for $CaCl_2$ in water,[17] which agreement further substantiates the specific hydration assumed. It is of interest to note also that the concentration indicated by the depression 51.6° is 9.25 mol. ($51.6 \div 5.58 = 9.25$). This concentration has the same rela-

[17] Int. Crit. Tables, 4: 257, gives $-51°C.$

tionship to the concentration of hydrated solute, 2.645 mol., as does the initial weight of solvent (1000) to the final weight of anhydrous solute at saturation (286). Since the molecular weight of $CaCl_2$ is involved in the foregoing relationships it is obvious that these freezing-point depression measurements may serve as indices of the weights of calcium and chlorine, the assumed weights being at variance with those commonly observed. As previously noted the matter of combining weight can not be considered in this paper.

The order of agreements above noted with respect to the depression of the freezing-point appears to be in support of the initial assumptions of this paper and the considerations developed through a study of electrical conductivity in relation to them. They appear sufficient, moreover, to warrant a more extended consideration of freezing-point measurements of concentrated solutions, but further studies can not be given space here.

BOILING-POINT ELEVATION AS AN INDEX OF HYDRATION IN THE LIGHTER ELEMENT IONS

The elevation of the boiling-point of any solvent by a solute is commonly considered as proportional to the number of molecules of solute present in a given weight of a solvent. For example, a molecular weight of a solute in grams when added to a liter of water in general raises the boiling-point 0.52°C.,—provided there is no ionization. An increase in the observed elevation over the expected one is interpreted as an index of ionization.

In the foregoing considerations of electrical conductivity and freezing-point depression in relation to an assumed hydration and change in weight, complete ionization of such electrolytes of KCl, LiCl, $CaCl_2$, etc., has been suggested at all concentrations. Consequently on the above basis we would expect that a gram-molecular-weight of KCl, for example, dissolved in a liter of water, would raise the boiling point twice the unit molecular amount, or 1.04°C. Similarly we would expect that a three-ion electrolyte, as $CaCl_2$, would raise the boiling-point $3 \times .52°$, or 1.56°, while $AlCl_3$ as a four-ion electrolyte would raise the boiling-point 2.08°C.

On the foregoing bases we may compare the calculated and observed elevations of the boiling-point characterizing solutions of electrolytes involving ions of the lighter elements as follows:[18]

[18] References for observed values: KCl, NaCl, Jablczyński and Kon, Jour. Chem. Soc., London 123: 2953, 1923. $CaCl_2$, Baker and Waite, Chem. and Metallurgical Engineering 25: 1174, 1921. Mg Cl_2, Kahlenberg, L., Jour. of Physical Chem. 5: 366, 1901. Li Cl, Biltz, Zeit. fur physik. Chemie, 40: 208, 1902.

208

TWO-ION ELECTROLYTES

Calculated KCl: Assuming for an approximation that one mole in 1000 parts by weight effects a unit elevation of .52°C., we have KCl, K^+ = 40, mol. wt. hyd. = 94, Cl^- = 32, mol. wt. hyd. = 158, 94 + 158 = 252, total mol. wt. 2 ions, 2 × .52° = 1.04°; *Observed KCl:* .8842 m. = .824°, .8842 × 252 = 223. *Calculated NaCl:* Na^+ = 24, mol. wt. hyd. = 222, Cl^- = 32, mol. wt. hyd. = 158, 222 + 158 = 380, total mol. wt. 2 ions, 2 × .52° = 1.04°, *Observed NaCl:* .9208 m. = .888°, .9208 × 380 = 350. *Calculated LiCl:* Li^+ = 8, mol. wt.

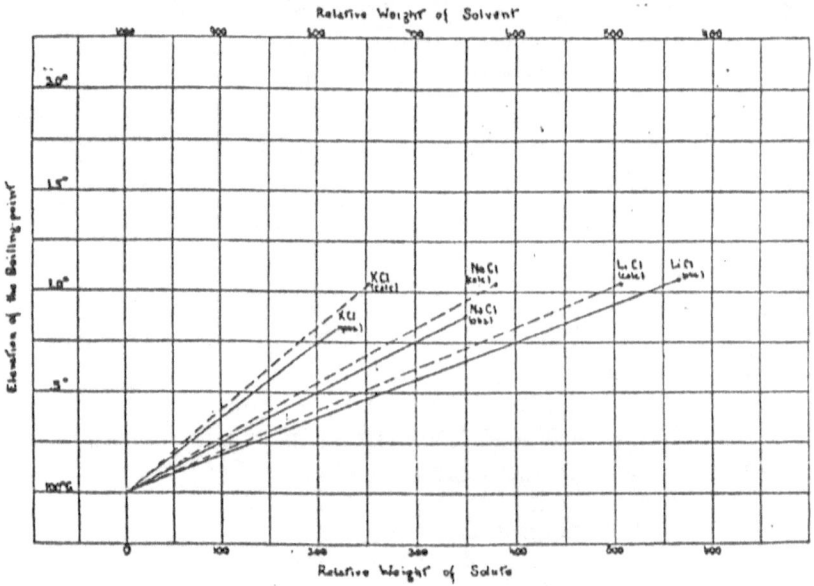

Fig. 5. Observed and calculated elevations of the boiling-point for some two-ion electrolytes.

hyd. = 350, Cl^- = 32, mol. wt. hyd. = 158, 350 + 158 = 508, total mol. wt. hyd. 2 ions, 2 × .52° = 1.04°, *Observed LiCl:* 1.05 m. LiCl gives an elevation of the boiling point of 1.063°C. *Observed Weight* LiCl = 42.48, *Calculated Weight* = 40, 42.48 ÷ 40 = 1.062, 1.062 × 1.05 = 1.115, 1.115 × 508 = 566.

THREE-ION ELECTROLYTES

Calculated CaCl₂: Ca^{++} = 44, mol. wt. hyd. = 62, Cl^- = 32, mol. wt. hyd. = 158, Cl^- = 32, mol. wt. hyd. = 158, 62 + 158 + 158 = 378, total mol. wt. 3 ions, 3 × .52° = 1.56°. *Observed CaCl₂:* 10 gms. CaCl₂ added to 100 gms. H₂O, gives boiling point of 101.3°C. Same as 100 gms. added to 1 liter H_2O, 100 ÷ 1100 = .091 or 91 parts per thousand.   91 ÷ 108 (theo. mol. wt. anhydrous $CaCl_2$) = 318. Elevation effected: 1.3°C.   *Calculated $MgCl_2$*: Mg^{++} = 28, mol. wt. hyd. = 190, Cl^- = 32, mol. wt. hyd. = 158, Cl^- = 32, mol. wt. hyd. = 158, 190 + 158 + 158 = 506, total mol. 3 ions, 3 × .52° = 1.56°.   *Observed $MgCl_2$*: 9.156 gms. added to 100 gms. water raised boiling point 1.351°C.   Equivalent to 91.56 gms. added to 1000 gms. water.   91.56 : 1091.56 :: x : 1000, x = 73.9 (equivalent to 83.9 gms. per 1000 gms. solution), 83.9 ÷ 92 (mol. wt. anhydrous $MgCl_2$)

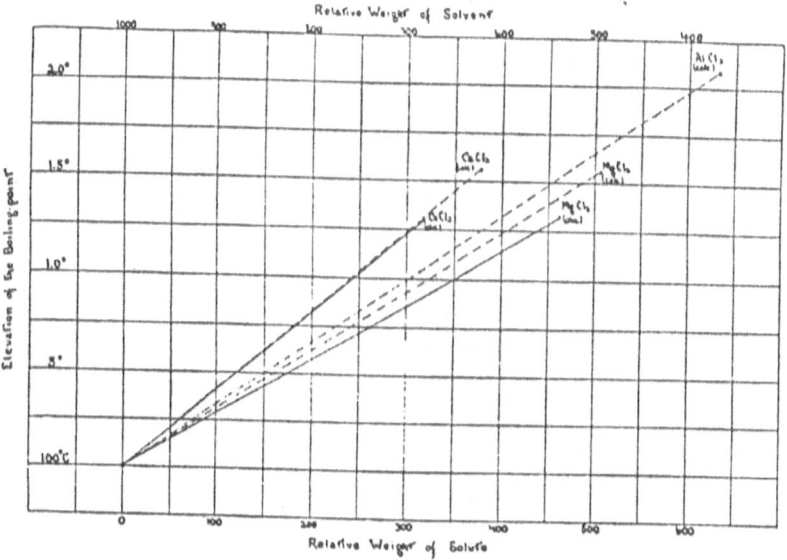

Fig. 6. Observed and calculated elevations of the boiling-point for some electrolytes of more than two ions.

= .912, .912 × 506 (mol. wt. hyd. $MgCl_2$) = 461.5, relative wt. of hydrated solute.

FOUR-ION ELECTROLYTES

Calculated $AlCl_3$: Al^{+++} = 32, mol. wt. hyd. = 158, Cl^- = 32, mol. wt. hyd. = 158, Cl^- = 32, mol. wt. hyd. = 158, Cl^- = 32, mol. wt. hyd. = 158, 158 (Al^{+++}) + 158 (Cl^-) + 158 (Cl^-) + 158 (Cl^-) = 632, 4 ions, 4 × .52° = 2.08°.   *Observed $AlCl_3$*: Corresponding observed values are not readily available at this writing.

An examination of the graphs in Figures 5 and 6 indicates that the calculated and observed values are in substantial agreement, and the boiling-point measurements to the extent of the agreements thus become subject to interpretation as indicating complete ionization at all concentrations. The data thus appear to support the suggestions of electrical conductivity and freezing-point depression in this regard.

With the addition of more and more solute to a solvent the boiling-point is raised higher and higher. The ratios of solvent to solute brought about by such concentrations suggest that the assumed attraction of the solute for the solvent is gradually offset through the elevation of the temperature. In any case the measurement of boiling-point elevation at high concentrations becomes of interest in relation to the initial assumptions of this paper, but such data are not readily available at present. Until such measurements become available the elevation of the boiling-point appears to be a measurement which can supply only indirect evidence for hydration.

SUMMARY

In the foregoing pages an inquiry has been made into the hydration of the solute ions of the lighter elements. Two initial assumptions were made (1) an inverse integral relationship between the anhydrous weight of a solute ion and the degree of its hydration and (2) an orderly change in weight accompanying ionization. Many observed measurements, involving electrical conductivity, freezing-point depression and boiling-point elevation have been noted as subject to a uniform interpretation on the basis of these assumptions. The order of agreement attained appears to warrant the extension of the inquiry to other ions,—a study which will be reported in a subsequent paper.

**Appendix B – Flint, L.H., *Journal of the Washington Academy of Science*
Vol. 22, No. 8, April 19, 1932, 211-217:
"The hydration of the solute ions of the heavier elements."**

CHEMISTRY.—*The hydration of the solute ions of the heavier elements.*[1]
L. H. FLINT. Bureau of Plant Industry. (Communicated by
G. N. COLLINS.)

In the first paper[2] of a proposed series dealing with the hydration of
solute ions it has been shown that an assumed inverse integral system
of hydration derived through the extension of Graham's Law to
solutions appears to characterize the stable element-ions of the first
quarter of the periodic system as judged by three aspects of solution
phenomena,—electrical conductivity, freezing-point depression and
boiling-point elevation. In the system as therein developed the
weight of the element appeared to be modified incident to ionization;
the respective hydration seemed entirely dependent upon the weight;
the hydrating water molecules were considered as uniting with the
atoms to form molecular ions of solute distinct from the solvent; and
complete ionization was indicated at all concentrations.

In studying the hydration characteristics of the heavier element
ions the most natural suggestion growing out of these previous considerations is to project an analogous system of inverse integral
hydration throughout the periodic system of elements. In accordance
with this suggestion values corresponding to the treatments given for
the lighter elements in Table 1 of the first paper have been worked out
and grouped herewith in Table 1,—each section comprising the elements
of a quarter of the periodic system. As was the case in connection
with the lighter elements of the first quarter, many of the elements
included are not known as stable ions in aqueous solutions. However,
since the assumed hydration appears to be conditioned by weight as
modified by ionization, the full set of values seems desirable as a
reference.

In the above table the assumed hydration in terms of water molecules per ion is given in the fifth column. The calculated weight
values for the unhydrated and hydrated states are given in the third
and seventh columns respectively, while the velocities corresponding
to these weight values are given in the fourth and eighth columns.

[1] Received February 24, 1932.
[2] *The hydration of the solute ions of the lighter elements.* This JOURNAL 22: 97-119.
1932.

TABLE 1.—Weight, Hydration and Velocity Values for a Postulated Inverse Integral Hydration System as applied to the Heavier Elements

A.N.	E	Assumed as W. 2 x A.N.	V_1	Postulated Number of Water Molecules	Mol. Wt. Water of Hydration	Mol. Wt. Hydrated Molecule	V_1
23	V	46	1475	23	414	460	466
24	Cr	48	1443	22	396	444	475
25	Mn	50	1414	21	378	428	484
26	Fe	52	1387	20	360	412	494
27	Co	54	1360	19	342	396	502
28	Ni	56	1336	18	324	380	513
29	Cu	58	1313	17	306	364	524
30	Zn	60	1290	16	288	348	536
31	Ga	62	1270	15	270	332	549
32	Ge	64	1250	14	252	316	563
33	As	66	1230	13	234	300	578
34	Se	68	1212	12	216	284	594
35	Br	70	1195	11	198	268	611
36	Kr	72	1178	10	180	252	630
37	Rb	74	1162	9	162	236	651
38	Sr	76	1147	8	144	220	675
39	Y	78	1132	7	126	204	700
40	Zr	80	1118	6	108	188	729
41	Cb	82	1105	5	90	172	763
42	Mo	84	1091	4	72	156	800
43	Ma	86	1078	3	54	140	846
44	Ru	88	1065	2	36	124	899
45	Rh	90	1054	1	18	108	963
46	Pd	92	1042	0	0	92	1042
46	Pd	92	1042	23	414	506	445
47	Ag	94	1031	22	396	490	452
48	Cd	96	1020	21	378	474	460
49	In	98	1010	20	360	458	467
50	Sn	100	1000	19	342	442	476
51	Sb	102	992	18	324	426	485
52	Te	104	982	17	306	410	494
53	I	106	972	16	288	394	502
54	Xe	108	963	15	270	378	514
55	Cs	110	954	14	252	362	526
56	Ba	112	945	13	234	346	538
57	La	114	936	12	216	330	551
58	Ce	116	929	11	198	314	565
59	Pr	118	922	10	180	298	580
60	Nd	120	914	9	162	282	595
61	Il	122	906	8	144	266	613
62	Sm	124	898	7	126	250	631
63	Eu	126	891	6	108	234	654
64	Gd	128	884	5	90	218	678
65	Tb	130	878	4	72	202	704
66	Dy	132	871	3	54	186	733

TABLE 1.—Weight, Hydration and Velocity Values for a Postulated Inverse. Integral Hydration System as applied to the Heavier Elements—*Concluded*

A.N.	E	Assumed as W. 2 x A.N.	V_1	Postulated Number of Water Molecules	Mol. Wt. Water of Hydration	Mol. Wt. Hydrated Molecule	V_2
67	Ho	134	864	2	36	170	768
68	Er	136	858	1	18	154	806
69	Tm	138	851	0	0	138	851
69	Tm	138	851	23	414	552	426
70	Yb	140	845	22	396	536	432
71	Lu	142	839	21	378	520	439
72	Hf	144	833	20	360	504	445
73	Ta	146	828	19	342	488	453
74	W	148	822	18	324	472	460
75	Re	150	817	17	306	456	468
76	Os	152	811	16	288	440	477
77	Ir	154	806	15	270	424	486
78	Pt	156	801	14	252	408	495
79	Au	158	796	13	234	392	505
80	Hg	160	790	12	216	376	516
81	Tl	162	786	11	198	360	527
82	Pb	164	781	10	180	344	539
83	Bi	166	776	9	162	328	552
84	Po	168	772	8	144	312	566
85		170	768	7	126	296	581
86	Rn	172	763	6	108	280	598
87		174	758	5	90	264	615
88	Ra	176	754	4	72	248	635
89	Ac	178	750	3	54	232	657
90	Th	180	746	2	36	216	680
91	Pa	182	741	1	18	200	707
92	U	184	737	0	0	184	737

The procedure corresponds precisely with that employed in Table 1 of the previous paper, and represents the extension of Graham's Law of the Diffusion of Gases to solute ions considered as of two potential states (1) hydrated and (2) unhydrated.

In examining evidence for the validity of the system as applied to the lighter element ions it was indicated that the apparent rate of decrease in specific molecular electrical conductivity with increasing concentration was subject to interpretation as an index of the rate at which the solvent becomes modified by the hydrating solute. The apparent rate of decrease thus constituted a convenient measure of hydration, and in the present paper this method of inquiry will be used exclusively in the examination of evidence for the validity of the system as applied to the heavier element ions and represented in Table 1.

With regard to the correspondence of the calculated and observed bases it may be noted that the concentrations used in the taking of measurements of electrical conductivity characteristically involve solutions made up to 1000 cc. rather than the 1000 grams involved in prediction. In dilute solutions the volume method is practically identical with the weight method, but in concentrations as great as 1.0 molecular the differences between the two systems are appreciable. When these differences are enhanced by hydration, particularly with ions having a high hydration, the amount of free solvent present in a one-molecular solution based on volume becomes difficult to evaluate with great precision. Then, again, observed measurements involve the use of the observed combining weights of the elements. As previously noted, these depart from the combining weights suggested under the hydration and weight-change hypotheses, and in the heavier elements the departure becomes appreciable. It seems desirable to consider the validity of the suggested hydration before specifically considering the validity of observed combining weights,—but it will be apparent that in view of the divergent weight bases represented in calculated and observed values a further source of discrepancy may be encountered. In general the use of the volume method for calculating concentration effects a dilution from the concentration calculated on the weight basis, while the use of the observed combining weights effects a concentration over the calculated basis. There is thus a tendency for these two factors to compensate each other with respect to the theoretical bases of calculation,—but under the circumstances an approximate agreement between observed and calculated values is all that may reasonably be anticipated, even were the hydration values known to be correct.

In examining observed conductivity measurements with reference to the values suggested in Table 1 we may assume a familiarity with the general treatments described in connection with the consideration of the lighter element-ions in the first paper.

Chromic Chloride, $CrCl_3$. The summation weight representing the solute in a 1.0 molecular solution of chromic chloride, $CrCl_3$, may be derived from Tables 1 of this and the previous paper, as follows:

$$Cr = 48, Cr^{+++} = 54, \text{ with } 19\ H_2O, \text{ mol. wt. hyd.} = 396$$
$$Cl = 34, Cl^- = 32, \text{ with } 7\ H_2O, \text{ mol. wt. hyd.} = 158$$
$$Cl = 34, Cl^- = 32, \text{ with } 7\ H_2O, \text{ mol. wt. hyd.} = 158$$
$$Cl = 34, Cl^- = 32, \text{ with } 7\ H_2O, \text{ mol. wt. hyd.} = \underline{158}$$
$$\text{Summation wt.} = \overline{870}$$

From this value the relative weight of solvent present may be derived as

$$1000 - 870 = 130, \text{ or } 13\% \text{ solvent}$$

Observed values for the specific molecular conductivity of $CrCl_3$ are available as follows: 1.0 mol. conc., 0°C.[3] $= 45.4$, .000244 mol. conc., 0°C.[4] $= 229.73$, $45.4 \div 229.73 = .1977$ or 19.77%. The observed conductivity at .000488 mol. conc. is 214.48 and a higher value than 229.73 is thus indicated for "zero" concentration. Considering this fact the order of agreement is such as to constitute evidence in substantiation of the postulated system of hydration.

Copper Chloride, $CuCl_2$. The summation weight representing the solute in a 1.0 molecular solution of copper chloride, $CuCl_2$, may be calculated from Tables 1 of this and the previous paper, as follows:

Cu $= 58$, $Cu^{++} = 62$, with 15 H_2O, mol. wt. hydrated $= 332$
Cl $= 34$, $Cl^- = 32$, with 7 H_2O, mol. wt. hydrated $= 158$
Cl $= 34$, $Cl^- = 32$, with 7 H_2O, mol. wt. hydrated $= 158$

Summation wt. $= 648$

From this value the relative weight of solvent present may be derived as

$$1000 - 648 = 352, \text{ or } 35.2\% \text{ solvent}$$

Observed values for the conductivity of $CuCl_2$ at 0°C. may be cited as follows:[5] volume $= 1.28$, conductivity $= 59.3$; volume .76, conductivity $= 48.22$; at zero concentration, conductivity $= 165$.[6] By interpolation, volume at 1.0 $=$ conductivity at 1.0 mol. conc. $= 53.3$. The relative conductivity may be derived as $53.3 \div 165 = .323$, or 32.3%. The order of agreement (35.2% as calculated, 32.3% observed) appears to constitute evidence that the Cu^{++} ion in a solution of copper chloride hydrates with 15 molecules of water as suggested by the postulated system.

Strontium Chloride, $SrCl_2$. The summation weight representing the solute in a 1.0 molecular solution of strontium chloride, $SrCl_2$, may be calculated from Tables 1 of this and the previous paper, as follows:

---

[3] Int. Crit. Tables, Vol. 6.
[4] Jones, H. C., Carn. Inst. Wash. Pub. #170, p. 62.
[5] Jones, H. C. and Getman, F. H. Am. Chem. Jour. 31: 327. 1904.
[6] Jones, H. C. and Bassett, H. P., on the other hand give 120.0 as the value (Carn. Inst. Wash. Publ. #180, p. 73). The use of this value as a base gives a somewhat higher figure than 32.3%.

$Sr = 76$, $Sr^{++} = 80$, with 6 H_2O, mol. wt. hydrated $= 188$

$Cl = 34$, $Cl^- = 32$, with 7 H_2O, mol. wt. hydrated $= 158$

$Cl = 34$, $Cl^- = 32$, with 7 H_2O, mol. wt. hydrated $= 158$

Summation wt. $= \overline{504}$

From this value the relative amount of solvent present may be derived as

$$1000 - 504 = 496, \text{ or } 49.6\% \text{ solvent}$$

Observed values for the conductivity of $SrCl_2$ in aqueous solutions at 0°C., may be cited as follows:[7] 1.0 mol. conc. $= 71.23$; .000488 mol. conc. $= 133.9$. The relative value at 1.0 mol. conc. may be derived as

$$71.23 \div 133.9 = .532, \text{ or } 53.2\%$$

The order of agreement (by calculation 49.6%, by observation 53.2%) appears to constitute evidence that the strontium ion, Sr^{++}, hydrates with 6 molecules of water as predicted by the system being tested.

Cadmium Chloride, CdCl₂. The summation weight representing the solute in a 1.0 molecular solution of cadmium chloride, $CdCl_2$, may be calculated from Tables 1 of this and the previous paper, as follows:

$Cd = 96$, $Cd^{++} = 100$, with 19 H_2O, mol. wt. hydrated $= 442$

$Cl = 34$, $Cl^- = 32$, with 7 H_2O, mol. wt. hydrated $= 158$

$Cl = 34$, $Cl^- = 32$, with 7 H_2O, mol. wt. hydrated $= 158$

Summation wt. $= \overline{758}$

From this value the relative weight of solvent may be derived as

$$1000 - 758 = 242, \text{ or } 24.2\% \text{ solvent}$$

Observed values for the conductivity of aqueous solutions of cadmium chloride, $CdCl_2$, at 18°C., may be cited as follows:[8] 1.0 mol. conc. $= 22.4$; .005 mol. conc. $= 91$. The relative value at 1.0 mol. conc. may be derived as $22.4 \div 91 = .246$, or 24.6%. The value at "zero" concentration would be somewhat higher than 91, yet the order of agreement (24.2% calculated, 24.6% observed) appears to indicate that in aqueous solutions of cadmium chloride the cadmium ion, Cd^{++}, hydrates with 19 water molecules as suggested by the extension of the hydration system into the third quarter of the periodic system.

There does not appear to be any electrolyte involving a representative element-ion of the group designated in Table 1, which is hydrated and yet soluble to the extent of a 1.0 molecular solution. On

[7] Jones, H. C., Carn. Inst. Wash. For 1.0 mol. conc. Pub. #180, p. 64; for .000488 mol. conc. Pub. #170, p. 39.

[8] Kohlrausch, F. und Holborn, L. *Leitvermögen der Elektrolyte,* p. 161.

this account the consideration of the validity of the tabulated values (except insofar as analogy may be invoked) must be deferred until a subsequent inquiry into the hydration of the nitrate ion, NO_3^-, and other molecular ions.

The order of agreement noted in the foregoing comparisons appears to constitute evidence for the assumed hydration of the involved ions. However, the further extension of the study leads to the suggestion that not all solute ions are hydrated,—a suggestion which will be considered in the next paper of this series.

Appendix C – Flint, L.H., *Journal of the Washington Academy of Science*
   Vol. 22, No. 9, May 4, 1932, 233-237:
   "Unhydrated solute element ions"

JOURNAL

OF THE

WASHINGTON ACADEMY OF SCIENCES

Vol. 22   May 4, 1932   No. 9

CHEMISTRY.—*Unhydrated solute element ions.*[1] L. H. Flint, Bureau of Plant Industry. (Communicated by G. N. Collins.)

In a consideration of the hydration of some solute element-ions as comprised in the first two papers of the present series[2] attention has been directed to a group of electrolytes whose components behaved rather consistently as hydrated ions in aqueous solution. For example, the chlorine ion, Cl^-, in solutions of the electrolytes KCl, NaCl, LiCl, $MgCl_2$, $CaCl_2$, $AlCl_3$, $CrCl_3$, $CuCl_2$, $SrCl_2$ and $CdCl_2$ as involved in measurements of electrical conductivity, seemed uniformly subject to characterization as having a hydration of seven water molecules, all the other ions also having the respective hydration values assigned them under the initial assumptions regarding hydration designated in Table 1, and interpreted through the assumption of change in weight with ionization. At this point we may examine some electrolytes which appear to give rise to unhydrated ions in aqueous solution.

Hydrochloric Acid, HCl. The relatively high velocity of the hydrogen ion, H^+, has frequently led to the conclusion that this ion is characteristically not hydrated. Various measurements of solution phenomena, moreover, have seemed to corroborate this conclusion. It will be of interest, therefore, to examine some observed relative velocities with respect to the assumptions of hydration and weight-change embodied in Table 1, and the above conclusion.

The ion-conductance of the ions K^+ and H^+ at 18°C. as cited by Creighton and Fink[3] and derived from observed electrical conductivities through the use of transference measurements, are as follows:

[1] Received March 28, 1932.

[2] This JOURNAL 22: 97–119 and 211–217, 1932. Herein are given the tables to which reference is made in this paper.

[3] Creighton, H. J. and Fink, C. J. *Electrochemistry.* Wiley and Sons, Vol. I, 1924, p. 134.

219

$K^+ = 64.5$, $H^+ = 313$. Referring now to Table 1 of the first paper we may derive the assigned value for the velocity of the ion K^+, considered as hydrated, in the following manner:

$K = 38$, $K^+ = 40 + 3 H_2O (3 \times 18) = 40 + 54 = 94$ (weight of hydrated ion, column 7). The velocity value corresponding with this weight is 1031 (column 8).

We may now consider the above ion-conductance values as relative velocities, since each ion carries the same charge, and determine the corresponding relative velocity of the H^+ ion on the tabulated scale by solving for x in the ratio

64.5 : 313 :: 1031 : x
obs. Vel.K^+ Obs. Vel.H^+ calc. Vel.K^+ (hydrated)

$x = 5000$.

Determining the indicated relative velocity of the H^+ ion as 5000 we note that there is no such figure comprised within the series of column eight, representing velocities of hydrated ions, but that the figure corresponds precisely with the figure representing the velocity of the hydrogen ion, H^+, considered as unhydrated, given in column four and derived as follows:

$H = 2$, $H^+ = 4$ *with no hydration*, $V_1 = 5000$

The agreement is in substantiation of the above-noted conclusion.

We may now proceed to compare the observed relative specific molecular conductivity of HCl at 1.0 molecular concentration with that which would be expected under the assumption that the H^+ ion was not hydrated

$H = 2$, $H^+ = 4$, with no hydration, ionic wt. = 4
$Cl = 34$, $Cl^- = 32$, with 7 H_2O, ionic wt. hyd. = 158
summation wt. = 162

$1000 - 162 = 838$, or 83.8% solvent.

Observed specific molecular electrical conductivities of HCl may be cited as follows:[4] 1.0 mol. = 199.85, "0" mol. = 236.92; 199.85 ÷ 236.92 = .844, or 84.4%.

The order of agreement is in further substantiation of the conclusion that the H^+ ion is characteristically not hydrated in aqueous solution.

Rubidium Chloride, RbCl. The summation weight suggested by Table 1 as representing the solute present in a solution of RbCl at 1.0 molecular concentration on the weight basis if the rubidium ion, Rb^+, does not hydrate may be calculated as follows:

[4] Jones, H. C. Carn. Inst. Wash. Pub. No. 180, 1913, p. 80.

220

$$Rb = 74, Rb^+ = 76; \text{ with no hydration, ionic wt. } = 76$$
$$Cl = \underline{34}, Cl^- = 32; \text{ with 7 } H_2O, \text{ mol. wt. hydrated } = \underline{158}$$
$$108 = \text{mol. wt., anhyd. summation weight } = \overline{234}$$

From this value the relative weight of solvent may be derived as
1000 − 234 = 766, or 76.6%.

The relative specific molecular electrical conductivity of RbCl at 1.0 molecular concentration may be derived from observed values at 18°C. as follows:[5] RbCl at .001 molecular concentration = 130.1; at 1.0 molecular concentration = 102; 102 ÷ 130.1 = .784, Rel. sp. mol. conductivity = 78.4%.

The apparent specific molecular conductivity at "zero" concentration may be presumed to be somewhat higher than that at .001 molecular concentration, with the relative value at 1.0 mol. somewhat lower. The agreement between the value calculated on the above basis (76.6%) and the observed value (78.4%) calculated from a somewhat low base, may be considered as evidence that the Rb+ ion in aqueous solutions of RbCl is not hydrated.

Caesium Chloride, CsCl. The summation weight suggested by Table 1 as representing the solute present in a solution of CsCl at 1.0 molecular concentration on the weight basis if the Cs+ ion does not hydrate may be calculated as follows:

$$Cs = 110, Cs^+ = 112; \text{ with no hydration, ionic wt. } = 112$$
$$Cl = 34, Cl^- = 32; \text{ with 7 } H_2O \text{ mol. wt. hydrated } = \underline{158}$$
$$\text{summation wt. } = \overline{270}$$

From this value the relative weight of solvent may be derived as
1000 − 270 = 730, or 73.0% solvent.

The relative specific molecular electrical conductivity of CsCl at 1.0 molecular concentration and 18°C. may be derived from observed values as follows:[6]

CsCl, 1.0 mol. conc. = 98.8, .0005 mol. conc. = 131.05; 98.8 ÷ 131.05 = .754. Relative specific molecular conductivity = 75.4%.

The apparent specific molecular conductivity at "zero" concentration may be presumed to be somewhat higher than that observed at .0005 mol. Under the circumstances the order of agreement between the calculated value (73.0%) and the value as observed (75.4%), appears to constitute evidence that the Cs+ ion in aqueous solutions of CsCl is not hydrated.

Although in the electrolytes HCl, RbCl and CsCl the positive ion

[a] Values cited are from Int. Crit. Tables, Vol. 6, p. 234.
[b] Previous citation.

221

has been indicated as unhydrated, there appears to be no reason why the absence of hydration may not characterize the negative ion.   On such an assumption we may examine further as follows:

Potassium Bromide, KBr.   The summation weight representing the solute present in a solution of potassium bromide, KBr, at 1.0 molecular concentration on the weight basis if the bromine ion, Br^-, does not hydrate may be calculated as follows:

K = 38, K^+ = 40, with 3 H_2O, mol. wt. hydrated   = 94
Br = 70, Br^- = 68, with no hydration, ionic weight   = 68

summation wt.   = 162

From this value the relative weight of solvent present may be derived as 1000 − 162 = 838, or 83.8% solvent.

The relative specific molecular conductivity of a 1.0 molecular solution of KBr, may be approximated from observed values at 0°C. as follows:[7] .5 molecular concentration = 65.82; .000976 molecular concentration = 79.23; 65.82 ÷ 79.23 = .831.   Relative specific molecular conductivity at .5 molecular concentration = 83.1%.   The corresponding value for 1.0 molecular concentration is not given in the reference cited, and would be somewhat less,—yet the indicated order of agreement appears to constitute evidence that the Br^- ion in aqueous solutions of KBr is not hydrated.

Potassium Iodide, KI.   The summation weight representing the solute present in a solution of potassium iodide, KI, at 1.0 molecular concentration on the weight basis if the iodine ion, I^-, does not hydrate may be calculated as follows:

K = 38, K^+ =   40, with 3 H_2O, mol. wt. hydrated = 94
I = 106, I^- = 104, with no hydration, ionic wt.   = 104

summation wt.   = 198

From this value the relative weight of solvent may be derived as
1000 − 198 = 802, or 80.2% solvent.

The relative specific molecular conductivity of a 1.0 molecular solution of KI may be derived from observed values at 18°C. as follows:[8] 1.0 molecular concentration = 96.8; .0005 molecular concentration = 121.2; 96.8 ÷ 121.2 = .799.   Relative specific molecular conductivity = 79.9%.

The agreement between the predicted value (80.2%) and the observed value (79.9%) is of an order to constitute evidence that the I^- ion in aqueous solutions of KI is not hydrated.

[7] Jones, H. C.   Carn. Inst. Wash. Pub. No. 170, 1912, p. 21.
[8] Jones, H. C. and Caldwell, B. P.   Am. Chem. Journ., May 1901.

The consideration of electrical conductivity measurements of aqueous solutions of HCl, RbCl, CsCl, KBr, and KI suggests that there may be hydrated ions in association with unhydrated ions, and that the unhydrated state may characterize either the positively-charged or the negatively-charged component.

It will now be of interest to note that a similar examination of the relative specific molecular conductivities of the electrolytes cadmium bromide, $CdBr_2$, and cadmium iodide, CdI_2, leads to the conclusion that in aqueous solutions of these salts all ions are hydrated. Yet we have heretofore noted that in association with potassium as KBr and KI the ions Br^- and I^- were indicated as not hydrated. Under the circumstances it appears that the presence or absence of the hydration of the Br^- and I^- ions may be a matter of association. Since by the precepts of the present inquiry hydration conditions the velocity of an ion it follows that we have come into variance with the Kohlrausch Law of the Independent Migration of Solute Ions, which holds velocity as independent of association.

In connection with the consideration of the hydration characteristics of inorganic and organic molecular ions in subsequent papers of this series it will be of interest to note from time to time further evidence that association may condition the presence or absence of hydration. Within the limits of a specified state, hydrated or unhydrated, the Law of Kohlrausch has been shown in this and in the foregoing papers to be applicable in substantial measure to concentrated solutions. Yet with the accession of additional evidence that the two states, hydrated and unhydrated, may characterize the ions of the same elements in different electrolytes, it would appear that the law would fail as a correct description of velocity relationships.

Errata:

p. 47 line 15, change "D_2" to "$(D_{1)}/2$"; line 21, change "with" to "within".

p. 91, new para. 2, line 3, change "$CuSO2.$" to "$CuSO4.$"

p. 94 spelling, para. 2, line 1, correct "researches"

p. 49, Table 8A, row 2, col. "Ma", change "7082" to "7071"; this error was carried over from Flint's 1932 Table 1, on p. 99.)
   Also, in row 2, change the listed Vh value from "379" to "396; and in row 2 He, change Vh from "365" to "378".

p. 99, para. 3 (after equation), line 2, change "M=7083" to "M=7071".

p. 109, Table 29, first row, change 7082 to 7071

p. 116, para 3, line 12, change "8 H O" to "8 H2O"
        line 13, change "O2+ and O2+" to "O2+ and O2-"

p. 123, correction, para 2, line 3, change "natural;" to "natural,"

p. 126, para 3, line 2, change "rediation" to "radiation".

p. 128, para. 2, line 3, change "seemd" to "seemed".

p. 145, para. 1, line 4, change "slso" to "also";
        line 6, change "valent 2-valent" to "2-valent"

p. 162, in equation following para 3 and in para 4, change 94 to 95.

p. 169, para 2, line 9, change "short" to "sort"

p. 171, para 4, line 5, change "substituated" to "substituted"

p. 187, para 1, line 12, change "intracellalar" to "intracellular"
        last para, line 8, change "out time" to "our time"

INDEX

www.ingramcontent.com/pod-product-compliance
Lightning Source LLC
Chambersburg PA
CBHW021423170526
45164CB00001B/66